THE SILENCE INSIDE DNA

What Ninety Eight Percent of Your Genome Has Been Trying to Tell Us

INGRID FALKENRATH

This book is a work of nonfiction. It is intended for general informational and educational purposes only and does not constitute medical advice. The scientific findings described within reflect the published research available at the time of writing. Readers with specific medical or genetic health concerns should consult a qualified healthcare professional or genetic counselor.

Every reasonable effort has been made to ensure the accuracy of the information presented in this book, drawing on peer reviewed scientific literature and other credible public sources. The author and copyright holder assume no responsibility for errors, inaccuracies, omissions, or any inconsistency herein.

Cover design and interior formatting by the author.

For the researchers who kept looking at the parts of the genome everyone else had already decided to stop looking at.

CONTENTS

Author's Note

This book is a work of popular science, written for curious general readers rather than specialists, and every specific factual claim within it, every percentage, every named researcher, every documented disease mechanism, is drawn from peer-reviewed scientific literature and other credible public sources rather than invented for narrative convenience. Where the underlying science remains genuinely unsettled or actively debated among working researchers, this book says so directly, because an honest account of an unfinished scientific story is more valuable, and ultimately more interesting, than a falsely tidy one.

Readers with a background in molecular biology may notice that certain technical simplifications were necessary to keep this account accessible to readers encountering these topics for the first time. Every effort has been made to ensure these simplifications never cross into inaccuracy. Readers without a scientific background should not feel any particular technical vocabulary is required before beginning. Every specialized term is introduced and explained plainly at the point it first becomes relevant, and a glossary at the end of the book offers a quick reference for anyone who wants to look a term up again later.

This is, above everything else, a book about revision: about how a field of science can hold a confident, reasonable, evidence-based belief for decades, and then patiently, rigorously, and without abandoning its basic commitment to evidence, revise that belief as better tools and better data become available. It happened once already, over the fifty years covered in the chapters ahead. It is very likely still happening now.

One structural note before beginning. The chapters that follow are organized roughly, though not strictly, in the order these discoveries actually unfolded, moving from the earliest debates over what non-coding DNA might be, through the specific molecular categories researchers have since identified within it, and finally to the newest generation of tools now being used to test what remains uncertain. Readers who prefer to jump directly to a specific topic, whether a particular disease, a particular molecule, or a particular scientist's story, should find the table of contents detailed enough to navigate directly there without losing the thread of the larger argument. Each chapter is written to stand reasonably well on its own, though the full picture, and the full pleasure of watching one discovery set up the next, comes through best by reading start to finish.

CHAPTER One

The Ninety Eight Percent Problem

It began, as so many revolutions do, with a number nobody wanted to believe.

In June of 2000, in the East Room of the White House, President Bill Clinton stood beside Francis Collins and Craig Venter to announce that scientists had produced a working draft of the human genome. Clinton called it the most important, most wondrous map ever produced by humankind. He was not exaggerating the achievement. He was, however, glossing over a discomfort that had already settled into the laboratories that did the work: the map showed far fewer genes than anyone had predicted, and it showed something else that took years to properly explain. Only a sliver of the three billion letters in that map appeared to do anything a textbook could name.

Before sequencing began, geneticists had made bets, sometimes literal ones, on how many genes a human being would need. Estimates ran as high as one hundred and forty thousand. The logic seemed sound. Humans are complicated. We grow eyes and immune systems and prefrontal cortices capable of composing symphonies and also capable of forgetting where we left the car keys. Surely such complexity demanded an enormous instruction set, something to rival or exceed the gene count of simpler creatures many times over.

The actual number, when it finally settled after a decade of refinement, was around twenty thousand protein-coding genes. A roundworm called Caenorhabditis elegans, a creature about a millimeter long with a nervous system of exactly three hundred and two neurons, carries roughly the same number. A grain of rice carries more. This was not a small embarrassment. It was a full-blown identity crisis for a field that had spent fifty years assuming gene count tracked with biological sophistication.

But the deeper surprise was not the gene count itself. It was what surrounded those genes. When researchers added up every stretch of DNA that actually gets translated into protein, the sequences responsible for building hemoglobin, insulin, collagen, antibodies, and every other protein your body manufactures, the total came to a little over one percent of the genome. Some estimates land closer to one point two percent for the coding sequence alone, and about three percent if you count the untranslated regions bookending each gene.

Either way, the conclusion is startling by any measure: more than ninety-seven percent of the DNA inside every cell in your body does not code for a single protein.

For decades, the scientific establishment had a name for this material, and it was not a flattering one. Junk. The term stuck hard enough that it appears in the title of the paper that first popularized it, a 1972 article by the geneticist Susumu Ohno called simply "So Much Junk DNA in Our Genome." Ohno was not being dismissive for the sake of it. He had a specific, technically reasonable argument, and we will spend an entire chapter examining it fairly, because his reasoning deserves more nuance than the word "junk" has since acquired in popular memory. But the label escaped its original scientific context and became something closer to a cultural verdict. If it doesn't make protein, it doesn't matter. That was the assumption embedded in three generations of biology education, and it is the assumption this book intends to dismantle, region by region, discovery by discovery, using nothing but documented science.

Here is the story in miniature, the one this entire book will unpack in detail: the ninety-eight percent of your genome that does not encode protein is not empty space. It is not evolutionary debris riding along for the sequencing ride. Large stretches of it are switches that determine which genes turn on, in which tissue, at which moment of development. Other stretches produce RNA molecules that never intend to become protein at all, molecules that instead fold into three-dimensional shapes and physically grab onto other molecules, silencing genes, activating genes, holding entire chromosomes in specific architectural postures inside the nucleus. Still other stretches are the fossilized wreckage of ancient viral infections, genetic parasites that colonized our ancestors' genomes millions of years ago and left behind sequences your cells have since repurposed for their own use, including, remarkably, sequences now essential to how the human placenta functions.

None of this was obvious in 2000. It has taken a quarter century of painstaking experimental work, involving thousands of scientists across dozens of countries, to build the case this book will present. The case did not arrive all at once, and it did not arrive without controversy. Some of the loudest disagreements in modern genomics have been fought over exactly how much of the non-coding genome truly matters, and those disagreements are not fully settled even now. This book will not pretend otherwise. Where scientists disagree, you will be told they disagree, and you will be given both sides plainly enough to form your own view.

What changed the conversation first was a project with an unglamorous acronym: ENCODE, short for the Encyclopedia of DNA Elements. Launched in 2003 by the National Human Genome Research Institute, the same agency that had overseen the sequencing effort, ENCODE set out to do something almost absurdly ambitious. Rather than asking what fraction of the genome codes for protein, a question already largely answered, ENCODE asked a much harder question: what does the rest of it actually do, biochemically, in living cells? The project eventually involved hundreds of researchers, generated data across dozens of human cell types, and in September 2012 published its central finding across more than thirty coordinated papers. At least eighty percent of the human genome, the consortium reported, showed evidence of specific biochemical activity in at least one cell type studied.

That eighty percent figure detonated inside the field. We will return to it at length in a later chapter, because the number itself is more complicated than headlines allowed, and the backlash against it, led by evolutionary biologists who pointed out that biochemical activity is not the same thing as evolutionary function, remains one of the sharpest scientific debates of the past fifteen years. But even the harshest critics of the ENCODE number did not dispute the underlying shift the project represented. The non-coding genome was no longer safe to ignore. It had a texture, a pattern, a set of biochemical fingerprints that demanded explanation.

This book exists because that explanation is now substantial enough to tell as a story, and because most readers have never heard it. You likely learned, if you learned anything about DNA in school, that genes are the meaningful part of your genetic code and that everything else is filler. That was the honest state of understanding when many of today's adults were educated. It is no longer the honest state of understanding today, and the gap between what the science now shows and what most people still believe is wide enough to walk through.

Consider a few of the discoveries that will unfold across these chapters. A woman's cells silence one of her two X chromosomes almost entirely at random early in development, a process essential for normal development, and the master switch controlling that silencing is not a protein-coding gene at all. It is a long strand of RNA called XIST that physically coats the chromosome it will shut down, spreading across it like a blanket, recruiting proteins that compress the DNA into a permanently inactive state. Consider that nearly half of your genome, by some estimates closer to fifty percent when every fragment and remnant is counted, traces back to transposable elements, sequences that originated as selfish, self-copying parasites, jumping from one location in the genome to another across millions of years of primate evolution, and that a woman named Barbara McClintock discovered this phenomenon decades before the tools existed to fully confirm it, working with nothing more exotic than corn kernels, patient observation, and a level of scientific stubbornness that eventually earned her an unshared Nobel Prize in 1983, more than three decades after she first proposed the idea to a skeptical, occasionally hostile scientific community.

Consider that specific expansions of short repeated sequences, three-letter or four-letter or five-letter blocks of DNA repeating far more times than they should, sit at the root of some of the most feared inherited diseases in medicine:

Huntington's disease, Fragile X syndrome, several forms of inherited ALS. These expansions occur almost entirely outside the classic coding sequence, and understanding them has required entirely new categories of genetic thinking, categories that did not exist when the phrase "junk DNA" first entered common use.

None of these are speculative or fringe findings. Every claim in this book traces to peer-reviewed research, much of it published in journals such as Nature, Science, Cell, and the Proceedings of the National Academy of Sciences, conducted at institutions including the National Institutes of Health, the Broad Institute, the Wellcome Sanger Institute, and university laboratories across five continents. Where a finding remains contested among working scientists, this book will say so directly rather than smoothing over the disagreement for a cleaner narrative.

Why does any of this matter to someone who is not a biologist? Because the story of the non-coding genome is quietly reshaping medicine in ways that will likely touch your own life or the life of someone you love. The overwhelming majority of genetic variants identified through genome-wide association studies, the large-scale efforts that link specific DNA differences to disease risk, fall outside protein-coding regions entirely. They sit in the regulatory landscape this book is about to explore. Cancer researchers now scan non-coding regions for mutations that switch on cancer-driving genes at the wrong time. Gene therapy designers borrow techniques originally developed to study transposable elements. Companies developing RNA-based medicines, the same broad category of biochemistry behind the mRNA vaccines that reached hundreds of millions of arms during the coronavirus pandemic, draw directly on decades of basic research into how non-coding RNA molecules function inside human cells.

The ninety-eight percent, in other words, is no longer a footnote to genetics. It has become one of its most active frontiers.

This book is organized to walk you through that frontier in roughly the order scientists themselves discovered it, though not always in strict chronological sequence, because clarity sometimes requires grouping related ideas together rather than following the calendar exactly. You will meet the scientists whose careers were shaped by these discoveries, some celebrated immediately, others ridiculed for years before vindication arrived. You will encounter the specific molecules and mechanisms now known to operate inside the supposedly silent majority of your genetic code, described in language built for a curious general reader rather than a laboratory colleague, but never simplified to the point of inaccuracy.

You will also encounter genuine scientific disagreement, presented honestly rather than resolved artificially for narrative comfort. The question of how much of the non-coding genome is truly functional, in the strict evolutionary sense of having been shaped by natural selection because it does something useful, remains one of the most contested numbers in modern biology. Estimates from comparative genomics, which look at which DNA sequences have been conserved across millions of years of evolution because mutations there tend to be harmful, suggest a functional fraction somewhere between eight and fifteen percent of the genome. That is a very different number from ENCODE's eighty percent biochemical activity figure, and the gap between those two numbers is not a rounding error. It reflects two legitimately different definitions of what "functional" should mean, and reasonable, well-credentialed scientists land on different sides of that definitional divide. This book will not pretend the debate is settled. It is not. What has changed, decisively and irreversibly, is the burden of proof. Fifty years ago, the default assumption was that non-coding DNA did nothing until proven otherwise. Today, the default assumption among working genome scientists has flipped. Non-coding sequence is now investigated on the working hypothesis that it might matter, because so much of it already has been shown to.

That reversal, quiet and technical as it may sound, represents one of the more significant shifts in how biologists understand the fundamental architecture of life. It touches how we think about human uniqueness, given how much of our extra non-coding sequence, compared to close relatives like chimpanzees, appears concentrated in regulatory regions rather than protein-coding genes themselves. It touches how we think about disease, given how many inherited conditions trace back to malfunctions in regions once assumed irrelevant. It touches how we think about evolution itself, given how large a role transposable elements appear to have played in reshaping the regulatory circuitry of the human lineage.

The chapters ahead will take these threads one at a time, slowly enough to actually understand them, but with enough momentum, I hope, that you will find yourself wanting to know what happens next the way you might with any well-told story, because this genuinely is one. It has false starts and vindicated outsiders, decades-long feuds between reasonable scientists, and discoveries that reorganized entire subfields within a matter of months. It has a woman who was laughed at for twenty years and then handed a Nobel Prize. It has viruses that infected our ancestors millions of years ago and left behind genetic tools our species now depends on to grow a placenta. It has molecules that do nothing you were ever taught molecules could do: RNA strands that never become protein but instead fold into precise three-dimensional shapes and physically hold chromosomes in place.

Somewhere inside that ninety-eight percent, in other words, is a genuinely gripping account of how the most intimate text in your body works, most of which has nothing to do with the small set of instructions you were taught to think of as "the genes." It is time to read the rest of the page.

CHAPTER Two

The Word That Was Not Quite True

Susumu Ohno did not set out to insult anyone's DNA.

When the Japanese American geneticist published his 1972 paper at a symposium on gene evolution, held at Brookhaven National Laboratory on Long Island, he was working through a genuinely difficult mathematical problem, one that had been nagging at population geneticists for over a decade. The problem is worth understanding on its own terms before we get to the phrase that made Ohno famous, because the phrase has long since outrun the argument that produced it, and the argument was more careful than its legacy suggests.

Here is the puzzle Ohno was chasing. Every time a cell divides, its DNA must be copied, and copying is not a perfectly faithful process. Mutations creep in. Most mutations that land inside a functional gene are neutral or mildly harmful, and a smaller fraction are severely harmful, capable of causing disease or reducing reproductive success. Population geneticists in the 1960s, particularly a British scientist named

J.B.S. Haldane and later the American Hermann Muller, had worked out that a species can only tolerate so much of what they called genetic load, the cumulative burden of harmful mutations across a population, before that burden starts dragging down average fitness and reproductive success. If every single base pair in a genome were functionally essential, meaning any mutation anywhere could be seriously damaging, then a species with billions of base pairs and a realistic mutation rate would accumulate more harmful mutations per generation than natural selection could plausibly weed out. The population would be crushed under its own mutational weight, a scenario evolutionary biologists still refer to as mutational meltdown.

Ohno looked at the human genome, or rather at the best estimates available in 1972 of its total size and its likely gene count, and ran the numbers. Something did not add up if every base pair mattered. The genome was simply too large, and the estimated mutation rate too high, for all of it to be under the kind of selective pressure that functional sequence requires. Ohno's proposed resolution was that most of the genome must therefore be free to accumulate mutations without consequence, because it was not doing anything selection needed to protect. He called this material junk DNA, borrowing a term that had circulated informally among a few geneticists before him, and the phrase appeared in his paper's very title.

It is worth pausing on what Ohno actually meant, because it differs meaningfully from how the phrase later got used. Ohno was not claiming that non-coding DNA was proven to be useless. He was making a population genetics argument about mutational tolerance, essentially arguing that most of the genome could not be under strong selective constraint given the numbers involved, and therefore probably was not doing anything selection would protect. That is a more careful claim than "this DNA is worthless garbage," though the distinction was quickly lost in translation as the idea spread beyond specialist journals.

A second geneticist deserves equal credit, or equal scrutiny, for cementing the concept in the public imagination: David Comings, whose 1972 paper "The Structure and Function of Chromatin" explored similar territory and helped popularize the idea that repetitive DNA sequences served no essential purpose. Comings leaned further into the idea that repetitive sequences specifically, the kind we will examine in detail in the chapters on Alu and LINE-1 elements, were likely nonfunctional byproducts of a genome's history rather than active participants in it.

By the mid-1970s, "junk DNA" had become genuinely useful shorthand inside the field, not because scientists had definitively proven the material did nothing, but because the term captured a reasonable working hypothesis given the evidence and mathematical constraints available at the time. Textbooks adopted it. Popular science writing adopted it even more enthusiastically, because "junk DNA" makes for a punchier sentence than "sequence of currently unknown but plausibly limited function subject to relaxed selective constraint." Richard Dawkins, in his 1976 book The Selfish Gene, offered an influential variation on the theme, suggesting that much of the genome might be more accurately understood as sequence that exists because it is good at getting itself copied, regardless of whether it benefits the organism carrying it. This idea, developed further in a pair of widely read 1980 papers in Nature by Leslie Orgel and Francis Crick, and separately by Ford Doolittle and Carmen Sapienza, introduced the concept of "selfish DNA," genetic sequence that persists across generations not because it helps the organism, but because it is simply effective at replicating itself within the genome, riding along generation after generation like a stowaway that provides no service to the ship but also does not sink it.

This selfish DNA framing turned out to be more prescient than the blunter "junk" label, because it correctly anticipated something that would take decades of subsequent research to fully document: much of the non-coding genome is not inert filler at all, but the accumulated residue of genetic elements that were, at some point in their evolutionary history, quite actively "selfish," meaning they existed because they were good at copying and inserting themselves throughout the genome. Transposable elements, which we will explore at length in the next two chapters, fit this description precisely. They are not junk in the sense of being useless byproducts. They are, or were, active genetic agents pursuing their own replication, and the fact that most of them have since been rendered inactive by accumulated mutations does not mean the genome around them stopped mattering. Quite the opposite, as it turns out: evolution has repeatedly repurposed these "selfish" sequences for the host organism's own benefit, a process biologists call molecular domestication or exaptation, which will occupy substantial space later in this book.

So where did the disconnect happen? How did a careful, mathematically grounded hypothesis about mutational load become a blanket dismissal of ninety-eight percent of the human genome?

Part of the answer is simply the mechanics of scientific communication. A precise, hedged claim made at a 1972 symposium does not travel through decades of textbooks, journalism, and casual scientific conversation while retaining its precision. Each retelling tends to strip away qualifications. "Much of the genome is probably not under strong selective constraint, based on population genetic modeling" becomes, a few retellings later, "most of your DNA is junk," which becomes, a few retellings after that, "scientists have proven that ninety-eight percent of your genome does nothing." That final version, the one many readers of this book likely absorbed at some point in their education, was never actually established by anyone. It was a distortion of a real, defensible, but far more limited argument.

The other part of the answer is that, for roughly three decades after Ohno's paper, the tools available to test his hypothesis experimentally simply did not exist. You cannot definitively determine whether a given stretch of

DNA does something biologically important by staring at it. You need ways to measure whether it is transcribed into RNA, whether proteins bind to it, whether removing it changes an organism's traits, whether its sequence has been conserved across evolutionary time in a pattern consistent with selective pressure. Through most of the 1970s, 1980s, and 1990s, these tools existed only in primitive, laborious, gene-by-gene forms. Testing the function of a single non-coding region might consume a graduate student's entire doctoral thesis. Testing millions of non-coding regions across the whole genome, the kind of systematic survey that would eventually challenge the junk DNA hypothesis at scale, was not remotely feasible until high-throughput sequencing technology matured in the 2000s.

This is an important point for understanding the shape of this book's story, so it is worth stating plainly: the junk DNA hypothesis was not overturned because scientists suddenly grew wiser or more open-minded. It was overturned, gradually and unevenly, because the technology to actually test it at scale did not exist until relatively recently, and once it did exist, the results it produced repeatedly and specifically contradicted the assumption that most non-coding sequence was functionally inert.

There were, even during the height of the junk DNA consensus, individual voices raising doubts, and their stories deserve more attention than they typically receive. Susumu Ohno himself, in later writing, acknowledged that some portion of what he had called junk would likely turn out to have functions not yet understood. He was not claiming permanent, universal irrelevance. He was describing a starting hypothesis appropriate to the evidence of 1972, and good scientists revise hypotheses when better evidence arrives. Barbara McClintock, whose extraordinary story occupies the next chapter, had already spent decades arguing, largely to a skeptical or indifferent audience, that transposable elements were active, regulatory, and biologically consequential rather than inert junk, though her specific arguments would not be broadly appreciated until long after she made them.

By the late 1990s and accelerating sharply through the 2000s, a specific and growing body of evidence began chipping away at the blanket junk DNA assumption from multiple directions simultaneously. Comparative genome sequencing across species revealed that certain non-coding regions were more evolutionarily conserved, meaning more resistant to accumulating mutations over millions of years, than the surrounding sequence, a pattern that strongly implies those regions are doing something selection actively protects, because if they truly did nothing, random mutation would have scrambled them just as freely as everything else nearby. Researchers cataloging RNA molecules produced by human cells found, to considerable surprise, that a much larger fraction of the genome was being transcribed into RNA than the fraction that actually coded for protein, raising an obvious and pressing question: what is all that extra RNA for? And a slowly accumulating set of individual case studies, gene by gene, mechanism by mechanism, began demonstrating specific, verifiable functions for specific non-coding sequences, functions ranging from gene regulation to chromosome structure to entirely new categories of biological activity nobody had previously catalogued.

By the time the ENCODE project published its landmark 2012 results, the ground had already been shifting under the junk DNA hypothesis for well over a decade. ENCODE did not single-handedly overturn the concept, despite how the popular press sometimes framed the story. It delivered the most systematic, comprehensive evidence yet assembled, across the largest scale yet attempted, that the non-coding genome was, in aggregate, biochemically active in ways that demanded serious explanation. The scientific debate that followed, which we will examine closely in a later chapter, was not really a debate about whether ENCODE's raw data was accurate. It was a debate about what that biochemical activity actually meant, and whether "biochemically active" was the same thing as "functionally important in the evolutionary sense." That distinction turns out to matter enormously, and untangling it will occupy a significant portion of this book.

For now, the important takeaway is simpler than the technical debate that followed it. The phrase "junk DNA" was never quite the confident, settled scientific verdict it came to represent in popular understanding. It began as a careful, mathematically grounded hypothesis appropriate to the evidence of its time, made by scientists who explicitly anticipated it would need revision as better tools and data arrived. The revision has arrived, in installments, over five decades, and it continues arriving today. The chapters that follow trace that revision in detail, starting with a scientist whose own work anticipated much of it decades before the rest of the field caught up, and whose story remains one of the great vindication narratives in the history of modern biology.

CHAPTER Three

Barbara McClintock and the Jumping Genes

CHAPTER THREE

Barbara McClintock and the Jumping Genes

She spent whole seasons walking the rows of a Long Island cornfield with a notebook, tracking kernels the way an astronomer tracks stars, and for the better part of two decades almost nobody outside a small circle of specialists took her seriously.

Barbara McClintock arrived at Cold Spring Harbor Laboratory in 1941, already an accomplished cytogeneticist with a reputation for meticulous, painstaking work on the chromosomes of maize. Corn was, and remains, an unusually good organism for this kind of research, because its kernels display visible color patterns that trace directly back to specific genetic events, letting a sharp-eyed researcher read genetic history off the surface of a single cob. McClintock had that eye. She could look at a kernel's pattern of colored and uncolored patches, streaks and speckles interrupting what should have been a uniform color, and infer things about the underlying chromosomes that most of her contemporaries could not.

What she noticed, beginning in the 1940s and developing over the following decade, was strange enough that it took her years of careful, repeated experimentation before she trusted the conclusion herself. Certain genetic elements in corn were not staying where they belonged. A gene controlling kernel color would sometimes appear to switch on or off in ways that could not be explained by ordinary inheritance. Tracking the pattern across generations, McClintock determined that specific segments of DNA were physically moving from one location on a chromosome to another, and that their new location could disrupt or restore the function of nearby genes depending on exactly where they landed.

She called these mobile segments "controlling elements," and she proposed that they represented an entirely new category of genetic behavior: DNA sequences capable of relocating within the genome, and capable, once relocated, of regulating the activity of neighboring genes. This was a genuinely radical claim for the early 1950s. The prevailing model of genetics at the time treated genes as fixed points on a chromosome, arranged like beads on a string, stable and immovable across generations except through the recognized mechanisms of mutation and recombination. McClintock was proposing that some genetic elements did not behave like beads at all. They behaved like something closer to trained operatives, capable of relocating and, once in a new position, capable of exerting regulatory control over whatever gene happened to sit nearby.

In 1951, McClintock presented these findings at the Cold Spring Harbor Symposium, one of the most important gatherings in genetics at the time. By most accounts, including her own later recollections, the reception was polite, confused, and largely unconvinced. The idea was too far outside the conceptual framework her colleagues were working within. Some geneticists reportedly found her presentation difficult to follow, not because her data was sloppy, but because the underlying concept required abandoning assumptions that had structured the field for decades. McClintock, by her own account, felt sufficiently discouraged by the response that she largely stopped presenting her transposable element research at major conferences for years afterward, though she continued the underlying experimental work at Cold Spring Harbor without interruption, publishing her findings in more specialized venues where they attracted a smaller, more attentive audience.

This period deserves honest treatment, because it is sometimes flattened into a simple story of a lone genius ignored by a closed-minded establishment, and the reality carries more texture than that. Some of McClintock's contemporaries did take her seriously and did attempt to incorporate her findings into their own thinking, though transposable elements remained a minority interest within genetics for years. Part of the difficulty was technical: the molecular tools required to directly visualize or sequence a transposable element, to prove definitively that a specific piece of DNA had physically relocated rather than being explained by some other, less exotic mechanism, simply did not exist in the 1950s. McClintock's evidence, however carefully constructed, remained indirect, built from patterns of inheritance rather than direct molecular observation. Skepticism toward an indirect, conceptually radical claim is not inherently unreasonable in science. It becomes a problem only when it hardens into permanent dismissal regardless of subsequent evidence, and for a significant stretch of scientific history, that is roughly what happened to McClintock's work.

The vindication, when it eventually arrived, came from an unexpected direction: bacteria, and then a rapidly expanding cast of other organisms, as molecular biology matured through the 1960s and 1970s and gained the tools to directly examine DNA sequence rather than infer genetic behavior from inherited traits alone. Researchers studying bacterial genetics identified mobile genetic elements of their own, sequences that could insert themselves into new locations and disrupt gene function when they did so, matching McClintock's decades-old description with striking precision. As sequencing technology improved through the 1970s and into the 1980s, transposable elements were identified across an enormous range of organisms, from yeast to fruit flies to, eventually, the human genome itself. The phenomenon McClintock had described in corn kernels turned out to be not a maize-specific quirk but a fundamental feature of genomes across nearly the entire tree of life.

In 1983, the Nobel Committee awarded Barbara McClintock the Nobel Prize in Physiology or Medicine, unshared, for her discovery of mobile genetic elements. She was eighty-one years old. The award came more than three decades after her original 1951 presentation, and it remains, to this day, one of the longest gaps in Nobel Prize history between a foundational discovery and its formal recognition. When journalists asked her afterward how it felt to be vindicated after so many years, McClintock, by most accounts a private and famously unsentimental person about her own achievements, reportedly downplayed the drama of the story, suggesting she had simply continued doing the work she found compelling regardless of how it was received, and had trusted that the underlying biology would eventually speak for itself once the right tools existed to confirm it.

What McClintock discovered, in modern terminology, are called transposable elements, or transposons, and they represent one of the single most important categories of sequence within the non-coding genome this book is devoted to explaining. Broadly, transposable elements fall into two major mechanistic categories. DNA transposons, sometimes called "cut and paste" elements, excise themselves from one location in the genome and insert directly into another, physically relocating rather than duplicating. Retrotransposons, the more abundant category in the human genome and the one that will occupy the next two chapters in detail, work differently: they are first transcribed into an RNA copy, which is then reverse-transcribed back into DNA and inserted at a new genomic location, meaning the original copy stays put while a new copy appears elsewhere, a "copy and paste" mechanism that allows retrotransposons to multiply across a genome over evolutionary time rather than simply redistributing existing copies.

The scale of this phenomenon in humans is difficult to overstate, and modern estimates place it considerably higher than most non-specialists would guess. Depending on how conservatively you count degraded and fragmentary copies, transposable elements and their remnants make up somewhere between forty-five and fifty percent of the entire human genome. Nearly half of the DNA in every one of your cells traces its origin back to elements that, at some point across the roughly eighty million years since the primate lineage diverged from other mammals, and in some cases stretching back much further, functioned as mobile, self-copying genetic sequences, precisely the kind of "controlling elements" McClintock first identified in corn kernels in the 1940s.

Most of these ancient transposable elements are no longer active. Over millions of years, the vast majority have accumulated enough mutations, through the ordinary process of genetic drift acting on sequence no longer under strong selective constraint, that they can no longer copy themselves or move to new locations. They persist as molecular fossils, recognizable by their characteristic sequence signatures but functionally inert as mobile elements. A much smaller subset, which the next two chapters will examine closely, particularly certain young LINE-1 elements, remain capable of active transposition even in modern humans, continuing to generate new genetic variation between individuals and, occasionally, contributing to disease when an insertion lands somewhere disruptive.

But "functionally inert as mobile elements" does not mean "functionally irrelevant to the genome," and this distinction sits at the heart of what this book is trying to convey. Even transposable elements that lost their ability to jump millions of years ago frequently retain sequence features, particularly binding sites for regulatory proteins, that the host genome has since incorporated into its own regulatory circuitry. This process, which researchers call exaptation or molecular domestication, describes how a sequence element that originated for one purpose, in this case a transposon's own self-replication, gets repurposed by evolution for an entirely different function serving the host organism rather than the transposon itself. We will see specific, well-documented examples of this throughout the coming chapters: transposon-derived sequences that now function as enhancers controlling nearby genes, transposon-derived proteins that the placenta now depends on for its structure and function, and transposon insertions that appear to have reshaped entire regulatory networks at key moments in primate and human evolutionary history.

McClintock's own view of genome organization, articulated well before molecular biology had the tools to confirm most of it, anticipated much of this later understanding with striking accuracy. She did not think of the genome as a fixed, static blueprint. She thought of it as a dynamic, responsive system, one in which mobile elements played an active regulatory role, sometimes activated in response to environmental or cellular stress, restructuring gene expression patterns in ways that could help an organism respond to changing conditions. This view, radical when she first proposed versions of it in the 1950s, aligns closely with how many genome biologists now understand transposable elements: not as static junk, and not merely as ancient parasites frozen mid-infection, but as a dynamic layer of genomic architecture that has been actively shaping gene regulation across the entire history of complex life, including, quite specifically, the genome you are carrying in every cell of your body right now.

The next two chapters take a close, detailed look at the two transposable element families that dominate the human genome numerically: Alu elements, the short, primate-specific sequences that alone account for roughly ten to eleven percent of your total DNA, and LINE-1 elements, the longer, older, and still occasionally active sequences that account for an even larger share. Both stories connect directly back to McClintock's cornfield, because both represent, at enormous scale, precisely the phenomenon she spent thirty years trying to convince a skeptical field to take seriously.

CHAPTER Four

Alu, the Hitchhiker

Roughly one out of every ten letters in your genetic code belongs to a single family of sequence, repeated more than a million times, none of it longer than about three hundred bases. If you printed the entire human genome as text, at ten letters per inch, and highlighted every occurrence of this one family in yellow, entire pages would glow.

This is Alu, and it is worth sitting with the scale for a moment before getting into the mechanism, because the scale alone tells you something important. This is not an obscure genetic curiosity tucked into some forgotten corner of chromosome nineteen. Alu sequences are woven through essentially every chromosome in the human genome, inserted into introns, wedged between genes, occasionally landing inside regulatory regions where their presence changes how nearby genes behave. Current estimates place the copy number at more than 1.1 million individual Alu elements, together accounting for approximately ten to eleven percent of total human DNA by length. For comparison, the entire collection of protein-coding sequence across your whole genome, every exon of every gene that builds every protein your body makes, totals a little over one percent. Alu elements alone outweigh the coding genome roughly tenfold.

The name itself comes from a technical detail of molecular biology history rather than anything more evocative. Alu elements contain a recognition site for a bacterial restriction enzyme called AluI, isolated from the bacterium Arthrobacter luteus, and researchers in the late 1970s, working out early methods for cutting and analyzing DNA fragments, named the newly identified repetitive sequence after the enzyme that had helped reveal its presence. It is an unglamorous origin story for a sequence family that would turn out to carry outsized biological consequences.

Alu belongs to a broader category of transposable elements called SINEs, short interspersed nuclear elements, distinguished from the longer LINE elements covered in the next chapter primarily by their size and by their dependence on other genetic machinery to move. Unlike LINE-1 elements, which encode their own complete transposition machinery, Alu elements are what molecular biologists call non-autonomous. They cannot copy and insert themselves using only their own sequence. Instead, they borrow the enzymatic machinery, specifically an enzyme called reverse transcriptase, produced by LINE-1 elements elsewhere in the genome, essentially hijacking a neighboring element's mobility toolkit for their own replication. It is a genuinely parasitic relationship, one genetic element exploiting the molecular equipment of another, and it has proven extraordinarily successful across primate evolutionary history.

Alu elements trace their origin to a small RNA molecule called 7SL RNA, part of a cellular structure called the signal recognition particle, which normally helps direct newly made proteins to the correct location within a cell. Somewhere around fifty five million years ago, in an early primate ancestor, a copy of this RNA sequence acquired the ability to be reverse-transcribed back into DNA and reinserted elsewhere in the genome, likely by co-opting LINE-1 machinery already present. That single molecular accident launched what would become one of the most successful genetic replication campaigns in mammalian history. Over the subsequent tens of millions of years, Alu elements copied and reinserted themselves across the primate genome at a scale that eventually produced well over a million distinct copies in the human lineage alone, each one a slightly diverged descendant of that original 7SL RNA sequence, distinguishable today by characteristic patterns of accumulated mutation that let researchers estimate roughly when each individual copy first inserted.

For years, the sheer number of Alu copies was treated as exhibit one in the case for junk DNA. A million nearly identical copies of a sequence with no known protein-coding function, scattered apparently at random throughout the genome, fit naturally into Ohno's framework of sequence too abundant and too repetitive to plausibly be under meaningful selective constraint. That framing has not held up well against subsequent evidence, and the accumulated case against it deserves careful walking through.

Begin with disease. Alu insertions are not evolutionarily frozen; a small number remain active enough to insert into new locations in modern humans, and when a new insertion happens to land inside or near a gene important for normal function, the consequences can be significant. Documented cases include Alu insertions disrupting the gene responsible for hemophilia, insertions contributing to certain forms of breast and colon cancer through disruption of tumor suppressor genes, and insertions linked to neurofibromatosis, a genetic disorder causing tumors to grow along nerve tissue. These are not hypothetical or statistical associations. They are specific, documented cases where researchers identified a new Alu insertion as the direct causal mutation responsible for a patient's disease, the same kind of direct causal evidence used to establish that any conventional gene mutation causes a given condition.

Beyond causing disease through direct insertion, Alu elements contribute to genetic disorders through a second, structurally distinct mechanism, one that depends specifically on their sheer abundance and repetitive similarity. Because more than a million nearly identical Alu copies are scattered throughout the genome, the cellular machinery responsible for repairing broken DNA strands can sometimes become confused during the repair process, mistaking one Alu copy for another similar copy located at a different chromosomal position entirely. This can trigger what geneticists call non-allelic homologous recombination, a repair error that results in large chunks of DNA being deleted or duplicated between the mismatched Alu sites. Researchers estimate that this Alu-mediated recombination mechanism accounts for a meaningful fraction, roughly half a percent by some published estimates, of newly arising human genetic diseases, a substantial figure given how many distinct mechanisms compete to cause inherited disease overall.

If the story stopped at disease, Alu elements would remain a compelling case study in genomic risk without necessarily overturning the junk DNA framework; after all, a sequence family can cause harm through structural accident without being functionally useful in any positive sense. But the evidence for genuinely beneficial, actively maintained Alu function has grown substantially over the past two decades, and this is where the story becomes more interesting than simple risk assessment.

Alu elements are transcribed by an enzyme called RNA polymerase III, the same cellular machinery responsible for producing several other essential classes of small RNA molecules, and this transcriptional activity is not uniform or random. Alu transcription increases specifically in response to cellular stress, including heat shock and viral infection, suggesting the cell actively regulates when and where Alu RNA gets produced rather than simply tolerating background transcriptional noise. Once produced, Alu-derived RNA sequences have been shown to interact directly with specific proteins involved in the cellular stress response, and accumulating research suggests these interactions play a genuine regulatory role in how cells respond to stressful conditions, rather than representing incidental byproducts of an overactive, poorly controlled transcriptional system.

Perhaps the most striking documented example of Alu-derived function involves a long non-coding RNA molecule called ANRIL, also known by its more formal designation CDKN2B-AS1, which is transcribed from a region of chromosome nine strongly associated with cardiovascular disease risk in large genome-wide association studies. Research has shown that Alu elements embedded within ANRIL and within its target genes are required for ANRIL to properly regulate those target genes, meaning the Alu sequence itself has become a functionally essential component of a documented human gene regulatory pathway with direct relevance to heart disease risk, a level of specific, mechanistically demonstrated function that goes considerably beyond simply calling a sequence "biochemically active."

Alu elements have also been repeatedly found embedded within the untranslated regions flanking protein-coding genes, the stretches of RNA sequence that surround the actual protein-coding portion of a messenger RNA molecule without being translated into protein themselves. In many documented cases, these embedded Alu sequences influence how stable the resulting messenger RNA molecule is, how efficiently it gets translated into protein, and in some cases whether it gets exported from the cell nucleus at all. Because so many human genes contain at least one Alu element within their untranslated regions, researchers now estimate that Alu-mediated regulation of messenger RNA behavior touches a substantial fraction of protein-coding genes across the genome, making Alu elements, cumulatively, one of the more significant contributors to how human genes are regulated after they are transcribed, a layer of biology that received essentially no serious attention until relatively recently.

There is also a broader evolutionary argument worth making explicitly, because it reframes the entire question in a way that the simple junk versus function framing tends to obscure. Alu elements are not merely present in the human genome. They are, in specific documented instances, exapted, meaning co-opted by the host genome for a function unrelated to their original replicative purpose, in ways that appear to have been actively shaped by natural selection because the resulting function benefits the organism. Alu-derived sequences have been identified functioning as alternative gene promoters, as exon sequences incorporated directly into mature messenger RNA molecules through a phenomenon researchers call exonization, and as binding sites for transcription factors that regulate nearby gene activity. Each of these represents a case where a sequence that originated as a selfish, self-replicating genetic parasite has been functionally integrated into the host genome's own regulatory architecture, occasionally to the point of becoming indispensable.

None of this erases the genuinely disruptive, occasionally disease-causing side of Alu biology described earlier in this chapter. Both things are true simultaneously, and holding both truths together is more accurate, and considerably more interesting, than forcing Alu elements into either a purely harmful or purely beneficial category. They are ancient genetic hitchhikers, numerous enough to occasionally cause serious structural damage through recombination errors, and numerous enough, and long-established enough within the genome, that evolution has also had ample opportunity across tens of millions of years to repeatedly repurpose individual copies for genuinely useful regulatory functions. Roughly one in every ten letters of your genetic code carries this dual legacy: molecular risk and molecular utility, inherited together, inseparably, from a single RNA molecule that first learned to copy itself into your ancestors' genome some fifty five million years ago.

CHAPTER Five

LINE 1 and the Ghosts That Still Move

Q: What is the only transposable element in the human genome still capable of moving entirely under its own power?

A: LINE-1, and the fact that it still can is stranger, and more consequential, than most people realize.

Let's start with the numbers, because they are larger than most readers expect. LINE-1, formally Long Interspersed Nuclear Element-1, accounts for somewhere between seventeen and twenty percent of the entire human genome depending on which counting method a given study uses, making it, by sequence length, the single largest identifiable category of DNA in your entire genetic code. There are more than five hundred thousand LINE-1 copies scattered across your chromosomes. Combined with Alu and other related SINE and LINE families discussed in the previous chapter, transposable elements derived from this broad retrotransposon lineage account for something approaching a third of total human DNA, generated almost entirely through the copy-and-paste replication mechanism these elements pioneered.

Here is what makes LINE-1 mechanistically distinct from Alu, and biologically more dramatic. Alu elements, as the previous chapter explained, are non-autonomous. They cannot move without borrowing molecular machinery produced by something else. A full-length, intact LINE-1 element carries no such dependency. It encodes its own complete transposition toolkit: two proteins, designated ORF1p and ORF2p, that together handle every step of the process a retrotransposon needs to successfully copy itself into a new genomic location. ORF1p binds to the LINE-1 RNA transcript and helps package it for transport. ORF2p carries two distinct enzymatic activities essential to the whole operation, reverse transcriptase, which converts the RNA copy back into DNA, and endonuclease, which cuts open the target DNA site where the new copy will be inserted. Together, this two-protein system allows a single intact LINE-1 element to autonomously copy itself and insert that copy elsewhere in the genome, entirely independent of any other genetic element, using only machinery encoded within its own roughly six-thousand-letter sequence.

Full-length, fully functional LINE-1 elements are actually quite rare within the enormous half-million-copy total. Most LINE-1 sequences in your genome are truncated, degraded fragments, remnants of ancient insertion events that lost portions of their sequence over evolutionary time, either through incomplete copying during the original insertion or through subsequent mutation, and these fragments are no longer capable of independent movement. Current research estimates that the average human genome carries somewhere between eighty and one hundred intact, potentially active LINE-1 copies capable of full autonomous retrotransposition, a small, specific subset carved out of the broader half-million-copy total, but a subset that remains biologically consequential precisely because it retains the ability to keep copying itself, generation after generation, inside living human cells today.

This ongoing activity has real, documented consequences for genetic diversity between individuals. New LINE-1 insertions continue to occur in the human germline, meaning they can be passed down to offspring, and researchers studying genetic variation across human populations have identified thousands of LINE-1 insertion sites that differ from one person to another, essentially a layer of genetic variation generated by ongoing transposition rather than by the more familiar mechanism of single-letter mutations. Some of these insertions land harmlessly in regions with no functional consequence. Others land inside or near genes, sometimes disrupting normal gene function badly enough to cause disease.

The list of documented LINE-1-caused genetic disorders is specific and well established in the medical literature, not speculative. A LINE-1 insertion disrupting the factor VIII gene has been identified as a direct cause of hemophilia A in specific documented patients. LINE-1 insertions have been identified as causal mutations in cases of Duchenne muscular dystrophy, in certain inherited forms of colon cancer through disruption of the APC tumor suppressor gene, and in specific documented cases of breast cancer. Each of these represents a case where researchers traced a patient's genetic disease directly back to a new, spontaneous LINE-1 insertion event disrupting a specific, previously functional gene, the same standard of causal evidence used to diagnose any conventional inherited genetic disorder.

Beyond inherited, germline insertions passed from parent to child, a separate and more recently appreciated phenomenon involves

LINE-1 activity occurring specifically within somatic cells, the non-reproductive cells that make up the vast majority of your body's tissue. Somatic LINE-1 activity, meaning new insertion events occurring in individual cells during a person's own lifetime rather than being inherited from a parent, has been documented at meaningfully high levels specifically within brain tissue. Research from laboratories studying human neural development, including influential work from the Salk Institute published across the 2010s, found that neurons in the human brain accumulate new LINE-1 insertions during development at rates substantially higher than most other tissue types, generating a degree of genetic mosaicism among individual neurons within a single person's brain that would have seemed implausible before this research existed. The full functional significance of this neuronal LINE-1 activity remains an active area of ongoing research, and this book will not overstate what is still genuinely unsettled, but the leading hypotheses under active investigation include possible roles in generating cellular diversity relevant to learning and memory, alongside separate lines of evidence connecting elevated or dysregulated LINE-1 activity to specific neurodegenerative conditions, including research linking LINE-1 dysregulation to the neuronal dysfunction observed in amyotrophic lateral sclerosis and frontotemporal dementia, particularly in cases connected to mutations in a gene called TDP-43, which normally helps keep LINE-1 activity properly suppressed within cells.

LINE-1 activity also appears connected to the biology of aging and cancer in ways that researchers are still actively working out, and it is worth being precise here about what is established versus what remains an active hypothesis. Cellular senescence, the state in which aging cells stop dividing but do not die, has been repeatedly associated with increased LINE-1 transcriptional activity and, in some studies, increased LINE-1 retrotransposition. One proposed mechanism connects this activity to inflammation associated with aging: LINE-1-derived DNA, once produced within an aging or senescent cell, can trigger an immune sensing pathway normally reserved for detecting viral DNA, since LINE-1's reverse transcriptase mechanism structurally resembles that of a retrovirus closely enough that the cell's own antiviral defenses sometimes respond to it as though it were an actual infection. Researchers investigating this connection, including notable published work exploring whether antiretroviral drugs, the same general class of medication used to treat HIV, might reduce age-associated inflammation by suppressing LINE-1 activity, represent one of the more genuinely novel translational research directions to emerge from basic LINE-1 biology in recent years, though it remains investigational rather than an established clinical treatment.

In cancer biology, elevated LINE-1 activity has been documented across a wide range of tumor types, and specific instances of LINE-1 insertions directly disrupting tumor suppressor genes or activating oncogenes have been identified in colorectal cancer, lung cancer, and other malignancies through large-scale cancer genome sequencing projects. Whether elevated LINE-1 activity in tumors is primarily a cause contributing to cancer development, a consequence of the broader genomic instability already present in cancer cells, or some combination of both operating simultaneously, remains a genuinely open and actively researched question, and honest treatment of the science requires acknowledging that uncertainty rather than resolving it artificially in either direction.

Given this substantial catalogue of disease associations, it would be reasonable to ask whether LINE-1 represents a purely destructive genomic element, an ongoing hazard the human genome has simply failed to fully suppress. The evidence does not support that conclusion either. Cells have evolved multiple, overlapping, actively maintained defense systems specifically dedicated to keeping LINE-1 activity in check, including DNA methylation that silences LINE-1 promoter regions, specialized proteins that degrade LINE-1 RNA before it can be reverse-transcribed, and the TDP-43 protein mentioned earlier, whose normal function includes suppressing LINE-1 expression. The existence of this elaborate, energetically costly, actively maintained suppression machinery is itself informative. Organisms do not typically evolve and maintain complex, multi-layered defense systems against genuinely inert, harmless sequence. The fact that human cells devote so much regulatory machinery to keeping LINE-1 activity contained tells us something important: LINE-1 retains enough capacity for genuine biological consequence, for better or worse, that evolution has found it worthwhile to keep the phenomenon under active, continuous control rather than simply ignoring it.

And, as with Alu elements in the previous chapter, LINE-1 sequence has also been repeatedly repurposed by the host genome for beneficial function through the same exaptation process described earlier. LINE-1-derived sequences have been identified functioning as alternative promoters driving the expression of specific human genes, and LINE-1 activity during very early embryonic development, in the first few cell divisions following fertilization, has been shown in multiple studies to play an active, apparently necessary regulatory role in normal early development, rather than representing purely disruptive background activity the embryo simply tolerates. Precisely how this developmental LINE-1 activity is controlled and what specific regulatory purpose it serves remains an area of active investigation, but its consistent, reproducible presence at this specific developmental stage, across multiple independent studies, strongly suggests genuine biological function rather than incidental noise.

There is a third major transposable element category worth introducing before this chapter closes, one distinct from both the LINE-1 elements just described and the Alu elements covered in the previous chapter, and it delivers on a promise made in this book's opening pages: an example of a genuine viral remnant that the human body now depends on for one of its most essential reproductive functions.

Endogenous retroviruses, commonly abbreviated ERVs, differ from LINE-1 and Alu in their evolutionary origin. Rather than arising from an internally duplicating cellular RNA molecule, as Alu did from 7SL RNA, endogenous retroviruses trace back to actual ancient viral infections, instances where a retrovirus, a category of virus that includes modern HIV among its living relatives, infected a germline cell of a distant ancestor and inserted a full copy of its viral genome into that ancestor's chromosomal DNA. If the infected individual survived and reproduced, that inserted viral genome was passed down to all of its descendants, generation after generation, gradually becoming a permanent, heritable fixture of the species' genome rather than a communicable infection. Endogenous retroviruses and their degraded remnants collectively account for approximately eight percent of total human DNA, a figure independently documented across multiple large-scale genomic surveys.

The overwhelming majority of these ancient viral genomes are, today, thoroughly broken. Millions of years of accumulated mutation have riddled most endogenous retrovirus sequences with disabling changes, rendering them incapable of producing infectious viral particles or causing any of the harm a modern viral infection might cause. Of an estimated eighteen distinct endogenous retrovirus lineages identifiable in the human genome, research has identified only sixteen that retain a full coding sequence for the viral envelope protein, and of those, only two have been found to retain the specific fusogenic activity, the capacity to trigger the fusion of two separate cell membranes into one, that made the original viral envelope protein useful to the virus that once carried it.

Those two exceptions carry a genuinely remarkable story. Rather than existing as inert viral wreckage, both have been captured and repurposed by the human genome for an essential biological function, and both have been given names reflecting that repurposed role: syncytin-1 and syncytin-2. Research beginning with a landmark 2000 study and substantially extended through the following two decades identified that syncytin-1, derived from a retrovirus belonging to what researchers classify as the HERV-W family, is expressed specifically and highly in human placental tissue, where it performs precisely the same membrane-fusing function it originally evolved to perform on behalf of an ancient infectious virus, except that the cells it now fuses are not host cells being invaded by a pathogen, but placental cells called cytotrophoblasts, which must fuse together to form the specialized multinucleated tissue layer called the syncytiotrophoblast, the layer directly responsible for the exchange of nutrients, gases, and waste between a pregnant person's bloodstream and a developing fetus.

Syncytin-2, derived independently from a separate endogenous retrovirus belonging to the HERV-FRD family, performs a closely related fusogenic function within the placenta and additionally appears to carry immunosuppressive properties inherited from its original viral function, properties that researchers believe may now contribute to preventing the pregnant person's immune system from rejecting the genetically distinct fetal tissue, a function directly analogous to how the ancestral virus's envelope protein would have originally helped it evade immune detection during active infection. Direct experimental confirmation of how essential these genes are came through knockout studies in mice, in which researchers genetically disabled the equivalent syncytin genes and observed resulting defects in placental cell fusion severe enough to compromise embryonic survival, providing direct causal proof, rather than mere correlation, that these virus-derived genes are now indispensable to normal mammalian reproduction.

The broader evolutionary picture that has emerged from subsequent comparative research is stranger still. Syncytin genes have been independently identified across a wide range of mammalian lineages, including separate syncytin genes in rodents, rabbits, carnivores, and ruminants, each traceable to a distinct, independent ancient retroviral infection specific to that lineage's evolutionary history. Because placental structure varies considerably across mammalian species, and because each lineage appears to have captured a different retrovirus for this purpose at a different point in evolutionary history, researchers have proposed that repeated, independent viral capture events of this general kind may have been a recurring, important driver of placental evolution across mammals broadly, with each newly captured retroviral envelope gene potentially contributing to the specific architecture of that lineage's placenta.

This is not a purely celebratory story, and this book's commitment to representing the science honestly requires noting the complications. Dysregulated syncytin-1 expression, meaning abnormal activity of this same gene outside its normal placental context or at abnormal levels within pregnancy, has been documented in association with several placenta-related pregnancy complications, including preeclampsia, and separately with a range of unrelated conditions including certain cancers, where its membrane-fusing activity outside the reproductive context it evolved for appears to contribute to tumor biology in ways researchers are still working out. The same protein that makes normal human pregnancy possible, in other words, can also contribute to disease when its expression is not properly restricted to the tissue and timing where it belongs, a pattern that should feel familiar by this point in the book: genuine, essential function and genuine capacity for harm, arising from the identical underlying molecular mechanism, depending entirely on the specific context in which that mechanism operates.

LINE-1, in other words, occupies a genuinely paradoxical position within the modern genome: the last fully autonomous molecular parasite still actively copying itself into human DNA, capable of causing serious disease when it lands somewhere disruptive, kept in check by an elaborate and costly cellular surveillance system, and, simultaneously, a sequence family that the genome has found ways to repeatedly harness for its own regulatory purposes across the long span of primate evolutionary history. It is neither simply junk nor simply beneficial. It is both, actively, at the same time, inside your cells, right now. And as the story of syncytin demonstrates, LINE-1 is not alone in this respect. Its more distant relatives among the endogenous retroviruses, remnants of infections that struck your ancestors millions of years before the human species existed, now sit at the very center of how every human pregnancy successfully proceeds.

CHAPTER Six

The Encyclopedia Nobody Quite Agreed On

There is a particular kind of scientific controversy that never quite gets resolved so much as it gets absorbed, its sharp edges worn down over years of continued argument until both sides more or less agree on the facts and simply keep disagreeing about what the facts mean. The ENCODE controversy is exactly that kind of dispute, and understanding it properly requires resisting the temptation to declare a clean winner, because the science itself does not offer one.

The Encyclopedia of DNA Elements project, universally known by its acronym ENCODE, began in September 2003, funded by the National Human Genome Research Institute with an explicit and enormous ambition: catalogue every functional element in the human genome, not merely the protein-coding genes that had already received the lion's share of scientific attention, but the entire regulatory and structural landscape surrounding them. The project unfolded in two major phases. A pilot phase, launched immediately in 2003, focused on just one percent of the genome, allowing the consortium to develop and refine its experimental methods on a manageable scale before attempting the full genome. That pilot phase concluded in 2007, publishing results that already hinted at the scale of biochemical activity researchers would eventually document across the complete genome.

The main production phase followed, eventually growing to include 442 researchers spread across research institutions worldwide, generating and analyzing data using dozens of distinct experimental techniques applied across more than one hundred forty different human cell types and tissue samples. This was, by any measure, one of the most ambitious coordinated data-generation efforts in the history of biology, comparable in scope and organizational complexity to the original Human Genome Project itself.

In September 2012, the consortium published its central findings simultaneously across thirty coordinated papers, released together in Nature and several other major journals, an unusual and deliberately choreographed publication event designed to convey the full scope of the project's conclusions all at once rather than allowing individual findings to trickle out piecemeal over subsequent months. The headline finding, the sentence that would end up quoted in newspapers worldwide and would eventually generate one of the most heated methodological debates in modern genomics, was this: the ENCODE consortium had assigned biochemical function to eighty point four percent of the human genome.

Read quickly, that sentence sounds like a direct, decisive refutation of the junk DNA hypothesis discussed earlier in this book. If eighty percent of the genome has biochemical function, then surely only a small remainder could plausibly be considered functionless junk. This is more or less how the finding was reported across much of the popular press, with headlines in respected outlets proclaiming the definitive end of junk DNA as a scientific concept. It is also, according to a substantial and vocal contingent of evolutionary biologists, a significant overstatement of what ENCODE's data actually demonstrated, and the reasoning behind that objection deserves careful, fair explanation, because it rests on a genuinely important distinction rather than mere pedantry.

The core of the disagreement concerns what the word "function" should mean in this context, and it turns out there are at least two legitimately different, scientifically defensible definitions in play, each associated with a different research tradition within genomics.

ENCODE's eighty point four percent figure was built on what researchers call a biochemical definition of function. Specifically, the consortium counted any genomic region as functional if it showed evidence of at least one specific biochemical activity in at least one of the cell types examined: being transcribed into RNA, being bound by a regulatory protein called a transcription factor, carrying a specific histone modification associated with active gene regulation, or existing within a region of accessible, open chromatin structure typically associated with active regulatory elements. By this biochemical, activity-based standard, the eighty point four percent figure is not seriously disputed as an accurate description of what ENCODE's experiments measured. The consortium's raw data collection and its core biochemical findings have held up well under subsequent scientific scrutiny and replication.

The objection raised by critics concerns whether biochemical activity, defined this way, is the right standard for claiming evolutionary or biological function in any meaningful sense. The most prominent and sustained critique came from Dan Graur, an evolutionary biologist at the

University of Houston, who published a sharply worded 2013 paper, co-authored with several colleagues, arguing that ENCODE's definition of function was so loose that it would classify almost any DNA sequence as functional given a sufficiently broad enough survey of biochemical assays across enough different cell types. Graur and his co-authors pointed out, with pointed rhetorical flourish, that this biochemical definition would technically classify a corpse as "functional" if it exhibited enough measurable biochemical activity, a deliberately provocative comparison meant to illustrate what they saw as a fundamental category error in how ENCODE had framed its central conclusion.

A second influential and more measured critique came from Ford Doolittle, the same evolutionary biologist whose earlier work on selfish DNA was discussed in an earlier chapter of this book, publishing his own 2013 paper in the Proceedings of the National Academy of Sciences. Doolittle's argument approached the problem from a different angle, built around what evolutionary biologists call the C-value paradox: the long-recognized observation that total genome size varies enormously across species in ways that bear no clear relationship to organismal complexity, with some amoeba species, for instance, carrying genomes many times larger than the human genome despite being single-celled organisms considerably simpler than a human being by any conventional measure. Doolittle proposed a thought experiment directly relevant to ENCODE's central claim: if a comparable biochemical survey were conducted across these much larger genomes, would researchers expect to find a similarly large percentage classified as biochemically functional, or would the absolute number of functional elements instead stay roughly constant while the percentage dropped as genome size increased around it. If the latter, Doolittle argued, this would suggest that much of the additional DNA in larger genomes serves a more general structural or bulk role rather than the specific, individually meaningful informational role ENCODE's language of function seemed to imply for each of the specific elements it catalogued, a distinction with real conceptual consequences for how the eighty percent figure should properly be interpreted.

Both Graur's and Doolittle's critiques, despite their different specific arguments, converged on a shared underlying concern, one that a 2018 follow up paper by Doolittle summarized under the heading of needing to stop being vague about what the word function actually means in genomics. Their shared worry was that

ENCODE's own published summary had, in their reading, implicitly encouraged a slide from a defensible, narrow claim, that eighty percent of the genome shows some biochemical activity in some cell type, toward a considerably less defensible, broader claim, that eighty percent of the genome is doing something specifically important for the organism in the way a conventional gene does. This book's account has tried, throughout, to keep these two claims clearly separated, in line with the caution both critics were urging.

The alternative standard favored by Graur and other evolutionary genomics researchers relies on comparative genomics and evolutionary conservation rather than direct biochemical measurement. The underlying logic runs as follows: if a specific DNA sequence performs an important biological function, then mutations disrupting that sequence will tend to be harmful to the organism carrying them, and natural selection will tend to eliminate those harmful mutations from the population over evolutionary time. This process, called purifying or negative selection, leaves a detectable signature: functionally important sequences tend to be more conserved, meaning more similar, across related species separated by millions of years of evolutionary divergence, than sequences under no selective constraint, which accumulate mutations freely and drift apart between species regardless of common ancestry. By this evolutionary conservation standard, estimates of the functional fraction of the human genome land considerably lower than ENCODE's biochemical figure, typically somewhere between eight and fifteen percent depending on the specific comparative genomic method and species comparisons used.

Both figures, the roughly eighty percent biochemical activity estimate and the roughly eight to fifteen percent evolutionary constraint estimate, have been independently reproduced by multiple research groups using multiple methods, and neither side of the dispute has seriously challenged the other's underlying data collection. This is worth emphasizing because it is easy to mistake the ENCODE controversy for a dispute about facts, when it is really a dispute about interpretation and definition. Both numbers are real, reproducible measurements of genuinely different things.

There is a reasonable, moderate position that has gained increasing acceptance among genomicists in the years since 2012, one that this book will adopt as its working framework going forward, while continuing to flag genuine remaining uncertainty wherever it exists. Biochemical activity, of the kind ENCODE measured, is a necessary but not sufficient condition for meaningful biological function. A sequence with zero biochemical activity, no transcription, no protein binding, no distinctive chromatin signature, is extremely unlikely to be functionally important, since it shows no detectable interaction with any of the cellular machinery through which genetic function would have to operate. But a sequence with measurable biochemical activity is not automatically functionally important either, because cellular machinery, particularly the enzymes and proteins responsible for transcription, can and does interact with DNA somewhat promiscuously, generating background biochemical signal at locations that may have no meaningful consequence for the organism's biology or fitness. The presence of biochemical activity, in other words, tells you where to look for possible function. It does not, by itself, prove that function exists.

Subsequent research through the 2010s and into the 2020s has increasingly focused on bridging this gap directly, using experimental techniques capable of testing function more rigorously than either biochemical signature alone or evolutionary conservation alone can provide. A particularly important development involves a class of techniques called massively parallel reporter assays, which allow researchers to directly test the regulatory activity of thousands of individual DNA sequences simultaneously within living cells, measuring whether each specific sequence actually drives gene expression rather than simply showing biochemical marks associated with regulatory activity. Separately, the CRISPR gene editing revolution, which began transforming molecular biology broadly around 2012, made it possible for the first time to systematically test the functional consequences of deleting or modifying specific non-coding sequences directly within living cells at a scale that was simply not feasible with older, more laborious genetic engineering techniques. A later chapter in this book examines these newer functional testing methods in detail, because they represent the current frontier of the field, moving the conversation beyond both ENCODE's original biochemical survey and the comparative genomics estimates that challenged it, toward direct, mechanistic proof of function, sequence by sequence, across the genome.

Where does this leave the honest reader trying to understand how much of the non-coding genome actually matters? Somewhere between the two headline numbers, almost certainly, though precisely where remains genuinely unsettled and is likely to remain a subject of active research and legitimate scientific disagreement for years to come. What can be said with confidence, and what this book has tried to demonstrate through specific, well-documented examples in the chapters already covered and the chapters still ahead, is that the functional fraction of the non-coding genome, however it ultimately gets defined and measured, is unambiguously larger than the "essentially nothing" assumption implicit in the popular understanding of junk DNA that prevailed for decades before ENCODE's 2012 publication. Whether the true figure eventually settles closer to ten percent or closer to fifty percent, the direction of travel across five decades of accumulating evidence has been consistently, unmistakably toward more function being discovered in territory once assumed empty, not less. That trajectory, more than any single contested percentage, is the real and lasting legacy of the ENCODE project, controversy and all.

CHAPTER Seven

Introns, the Pauses Between Words

Imagine trying to read a novel in which every third paragraph, seemingly at random, was written in a language you don't understand, sometimes running for a page, sometimes for twenty, before the story resumes exactly where it left off. Now imagine that your cells perform this exact feat, seamlessly, millions of times a day, every time they need to build a working protein from a gene.

This is, more or less, the situation facing every one of your protein-coding genes. Genes are not, in most cases, continuous, uninterrupted stretches of coding sequence. Instead, the vast majority of human genes are broken into segments called exons, which do contain the protein-coding instructions, separated by intervening stretches called introns, which do not. Before a gene's initial RNA transcript can be translated into a functioning protein, cellular machinery must precisely locate and remove every intron, splicing the remaining exons together into a continuous, readable sequence. Get this process wrong, even by a single misplaced letter, and the resulting protein can be garbled beyond function.

Introns were discovered in 1977, independently and almost simultaneously, by two research groups: one led by Phillip Sharp at MIT, working with adenovirus, and another led by Richard Roberts at Cold Spring Harbor Laboratory, examining the same viral system. Both teams found something that violated the prevailing assumption about how genes worked. Genetic information had been understood, since the foundational work establishing the genetic code in the 1960s, as a continuous, colinear correspondence between DNA sequence and protein sequence, meaning each successive stretch of DNA corresponded directly to the successive portion of protein it built, without interruption. Sharp's and Roberts's independent experiments showed instead that the RNA transcript produced from certain viral genes was considerably longer than the final, mature messenger RNA molecule that actually got translated into protein, and that the difference consisted of intervening sequence that was transcribed initially and then physically removed before translation occurred. This discovery, representing a genuine and fundamental revision to how genetic information flows from DNA to protein, earned Sharp and Roberts the joint 1993 Nobel Prize in Physiology or Medicine.

Introns turned out to be far more common, and far larger, in human genes than the earlier viral discoveries had suggested. Once the full human genome sequence became available, researchers found that introns collectively account for a substantially larger share of total gene length than the exons they interrupt. A typical human protein-coding gene may span tens of thousands of DNA letters from start to finish, yet the actual protein-coding exon sequence within that gene often totals only a few thousand letters, sometimes far fewer, with introns making up the considerable remaining bulk. Some individual introns extend across hundreds of thousands of base pairs, meaning a single intron can be dramatically larger than the entire protein-coding portion of the gene it interrupts.

For years, introns fit comfortably within the broader junk DNA framework discussed earlier in this book. If introns get removed entirely before a protein is built, and if their sequence shows no obvious pattern of conservation comparable to the exons they surround, then the straightforward inference was that intron sequence itself carried little independent biological importance beyond simply existing as material that needed to be spliced out. This inference, like the broader junk DNA hypothesis it belonged to, has required substantial revision as research accumulated evidence for genuine intron function operating through multiple, distinct mechanisms.

The most fundamental function introns enable is alternative splicing, a process through which a single gene can produce multiple distinct protein products by including or excluding different combinations of exons during the splicing process. Rather than always removing the same introns and retaining the same exons every time a given gene is transcribed, cells can vary which exons get included in the final, mature messenger RNA, depending on the specific tissue type, developmental stage, or cellular signaling context involved. Current estimates suggest that more than ninety percent of human multi-exon genes undergo some form of alternative splicing, meaning the large majority of your genes are capable of producing more than one distinct protein variant from the same underlying DNA sequence, simply by varying which exons get included or excluded during processing.

This has profound implications for how researchers now think about the relationship between gene count and biological complexity, a relationship this book raised as a puzzle back in its opening chapter. The human genome contains only around twenty thousand protein-coding genes, a number that struck many geneticists as surprisingly modest when the genome sequence was first completed. Alternative splicing offers a substantial part of the resolution to that puzzle. If a meaningful fraction of your twenty thousand genes can each produce several, sometimes dozens, of distinct protein variants through alternative splicing, then the effective diversity of proteins your genome can produce runs considerably higher than the raw gene count alone would suggest, potentially into the hundreds of thousands of distinct protein variants when every documented splicing variation across every gene is tallied. Introns, in other words, are not simply inert material removed and discarded. They are the structural feature that makes this entire additional layer of biological complexity possible in the first place, by creating the boundaries and splice sites that alternative splicing machinery uses to generate protein diversity.

The splicing process itself depends on specific sequence signals located within introns, not merely at their boundaries with adjacent exons but sometimes deep within the intron body itself. Short sequence motifs called splicing enhancers and splicing silencers, located within intron sequence, help recruit or block the cellular splicing machinery at specific locations, fine-tuning exactly how a given transcript gets processed. Mutations disrupting these intronic regulatory sequences, even though they occur entirely outside the protein-coding exons themselves, have been documented as direct causes of human genetic disease, including specific, well-characterized cases contributing to spinal muscular atrophy, certain forms of inherited cancer predisposition, and various other conditions where the underlying problem is not a change to the protein sequence itself but a change to how correctly that protein gets assembled from its component exons in the first place.

Introns also appear to play a documented role in a quality control mechanism called nonsense-mediated decay, through which cells detect and destroy messenger RNA molecules containing premature stop signals that would otherwise produce a truncated, likely nonfunctional protein. The molecular memory of where introns were originally spliced out, retained through protein complexes that remain associated with the mature messenger RNA even after splicing is complete, helps the cell's surveillance machinery determine whether a given stop signal appears in the correct, expected location or represents an error requiring the transcript's destruction before it can produce a defective protein. This system depends directly on the exon-intron structure of the original gene, meaning introns contribute functionally to genetic quality control even after their own sequence has already been physically removed from the transcript.

A further, more recently appreciated layer of intron function involves the introns themselves serving as a direct source of additional non-coding RNA molecules. Many microRNAs, the small regulatory RNA molecules explored in a later chapter of this book, are encoded within intron sequences and get processed out during the normal splicing process, meaning a single gene can simultaneously produce a functional protein from its exons and a functional regulatory microRNA from one of its introns, two entirely distinct biological products generated from the same underlying transcriptional event. Certain classes of small nucleolar RNAs, molecules involved in chemically modifying other RNA molecules within the cell, are similarly encoded within intron sequences of otherwise unrelated host genes, using the intron as a kind of convenient genomic address rather than any deeper functional relationship to the host gene's own protein product.

There remains genuine, ongoing scientific debate about how much of total intron sequence, as opposed to the specific regulatory elements and embedded RNA genes described above, carries meaningful function versus how much genuinely does behave as the kind of low-constraint filler sequence the original junk DNA framework anticipated. Introns are, on average, considerably less evolutionarily conserved across species than exons, consistent with at least large stretches of intron sequence being under comparatively relaxed selective constraint. This is one of the clearer examples in this book of a nuanced middle position being the most scientifically honest one: introns as a category are neither uniformly essential nor uniformly irrelevant. They contain specific, well-documented functional elements, splice site signals, splicing enhancers and silencers, embedded regulatory RNA genes, quality control machinery components, embedded within a broader sequence context that, in many regions, likely does tolerate mutation with fewer consequences. Sorting out precisely which portions of any given intron fall into which category remains an active area of genomic research, made considerably more tractable in recent years by the same high-throughput functional testing approaches, particularly

CRISPR-based screening, that a later chapter in this book will examine in detail.

What began, in 1977, as an unexpected wrinkle in how a single viral gene was processed has grown, across nearly five decades of subsequent research, into one of the central organizing features of how human genes actually work. The pauses between the words of your genetic code, it turns out, are not merely pauses. They are load-bearing structural elements of the sentence itself, shaping which proteins get made, in which combinations, in which tissues, at which moments of your life, in ways the discoverers of the first spliced viral gene in 1977 could scarcely have anticipated.

CHAPTER Eight

The Regulators in the Shadows

THE REGULATORS IN THE SHADOWS

Every cell in your body carries an identical copy of your genome. The neuron firing in your visual cortex right now, translating patterns of light into the experience of reading this sentence, contains the exact same DNA sequence as a skin cell on your elbow, a muscle fiber in your calf, or a cardiac cell keeping your heart beating at this very moment. And yet these cells look nothing alike, behave nothing alike, and perform entirely different biological functions. The obvious question, one that occupied developmental biologists for much of the twentieth century, is how identical genetic instructions produce such radically different outcomes.

The answer lies almost entirely outside the protein-coding genes themselves, in a category of non-coding DNA sequence collectively known as regulatory elements. This chapter covers three of the most important regulatory categories, promoters, enhancers, and silencers, along with a fourth topic that has only become experimentally accessible within the past fifteen years or so: the actual three-dimensional folding of DNA inside the cell nucleus, which turns out to matter every bit as much as the linear sequence of letters most genetics education focuses on almost exclusively.

Promoters sit at the most familiar end of this spectrum, and they are the regulatory element closest in concept to what a general reader might already associate with gene control. A promoter is a stretch of DNA sequence located immediately upstream of a gene's protein-coding sequence, serving as the docking site where RNA polymerase, the enzyme responsible for actually transcribing DNA into RNA, along with an entourage of additional proteins called general transcription factors, assembles before transcription of that particular gene can begin. Without a functioning promoter positioned correctly relative to a gene, that gene simply cannot be transcribed, regardless of how intact and functional its protein-coding sequence might otherwise be. Promoters were among the first regulatory elements identified and studied in detail, partly because their location, immediately adjacent to the genes they control, made them comparatively straightforward to find using the experimental techniques available even in the earlier decades of molecular biology.

Enhancers represent something considerably stranger, and their discovery required biologists to abandon an assumption that had seemed almost too obvious to question: that a piece of DNA regulating a particular gene ought to be located somewhere near that gene. Enhancers were first identified in 1981, through work on a virus called SV40, when researchers discovered a short DNA sequence capable of dramatically boosting the activity of a nearby gene's promoter, but with a peculiar and unexpected property. The enhancer sequence worked regardless of which orientation it was inserted in, and worked even when researchers experimentally moved it thousands of base pairs away from the gene it regulated, on either side. This was, at the time, a deeply counterintuitive finding. How could a regulatory sequence located far away, in either direction, and in either orientation, still exert precise, reliable control over a specific target gene's activity?

The answer, worked out over the subsequent decades through increasingly sophisticated experimental and imaging techniques, involves the physical, three-dimensional folding of DNA inside the cell nucleus. DNA is not stored inside your cells as a straight, extended line of sequence, the way a genetic map on a page might suggest. Instead, each chromosome is folded, compressed, and looped into an intricate three-dimensional architecture, and this folding brings sequences that appear far apart when measured along the linear DNA strand into close physical proximity within the actual folded structure. An enhancer situated tens of thousands, sometimes hundreds of thousands, of base pairs away from its target gene along the linear sequence can, through this folding, end up physically adjacent to that gene's promoter within the folded three-dimensional structure of the chromosome, close enough for proteins bound to the enhancer to directly contact and help activate the transcription machinery assembled at the promoter.

The human genome is now understood to contain a very large number of enhancer elements, considerably outnumbering the protein-coding genes they regulate. Estimates from ENCODE and subsequent large-scale regulatory mapping projects place the total number of candidate enhancer elements across the human genome somewhere in the range of several hundred thousand to over a million, depending on the specific criteria and cell types used to identify them, reflecting the reality that a single gene is often regulated by multiple distinct enhancers, active in different tissues, at different developmental stages, sometimes working in combination to produce precisely calibrated levels of gene expression appropriate to a very specific biological context.

Silencers work through a broadly analogous mechanism to enhancers, but in the opposite direction, reducing rather than boosting the activity of their target gene's promoter. Silencer elements bind specific repressor proteins that, once positioned near a gene's promoter through the same kind of three-dimensional chromatin folding that brings enhancers into contact with their targets, actively suppress transcription rather than promoting it. Silencers have received comparatively less popular attention than enhancers, but they are equally essential to normal development and gene regulation, since correctly turning genes off, at the correct time and in the correct tissue, is every bit as biologically important as correctly turning them on.

Understanding how these individual regulatory elements, promoters, enhancers, and silencers, actually find and contact their correct target genes within the crowded, three-dimensional space of the cell nucleus required an entirely new experimental technology, one that did not exist until relatively recently in the history of molecular biology. Traditional DNA sequencing tells researchers the linear order of genetic letters along a chromosome, but it says nothing directly about how that chromosome is physically folded inside a living cell. A technique called chromosome conformation capture, first published in 2002 and substantially refined and scaled up over the following decade into a genome-wide version called Hi-C, published in 2009 by a team including Erez Lieberman-Aiden and colleagues working across Harvard and MIT, finally gave researchers a practical way to directly measure which distant regions of the genome are physically close to one another inside the folded three-dimensional nucleus, by chemically freezing DNA in its folded state, cutting it into fragments, and measuring which fragment pairs end up stuck together.

What this technology revealed reshaped how genome biologists understand chromosome architecture. Rather than existing as a loosely tangled mass of DNA, chromosomes fold into a hierarchy of organized structures. At a broad scale, chromosomes occupy distinct, largely non-overlapping territories within the nucleus. Within each chromosome, DNA organizes into large domains, sometimes called topologically associating domains, within which sequences interact with each other frequently, while interactions across domain boundaries are comparatively rare, as though each domain forms something like an insulated neighborhood within the broader chromosome. Within these domains, specific loops bring individual enhancers into direct physical contact with their target gene promoters, often mediated by a protein complex called cohesin working in coordination with a DNA-binding protein called CTCF, which helps anchor the boundaries of these structural domains and regulate which loops can and cannot form.

This architectural picture has direct, clinically documented consequences for human disease, because disrupting the boundaries between these structural domains can cause an enhancer that normally regulates one gene to instead come into inappropriate physical contact with an entirely different gene located within a neighboring domain, activating that gene incorrectly, in the wrong tissue or at the wrong time, with serious developmental consequences. Documented cases include specific limb malformation disorders traced directly to structural mutations disrupting domain boundaries near genes in the developmentally critical WNT and SHH signaling pathways, where the underlying problem is not a mutation to the affected gene's own protein-coding sequence at all, but a structural rearrangement that allows an enhancer normally dedicated to a different, neighboring gene to inappropriately activate the disease-associated gene instead. These cases represent some of the clearest available evidence that three-dimensional genome architecture is not simply an interesting structural curiosity, but a functionally essential layer of gene regulation whose disruption can directly cause serious congenital disease, entirely independent of any mutation within the affected gene's own coding sequence.

Before turning to how this regulatory system builds different cell types, it is worth introducing one further discovery that connects enhancer biology directly to the question of what makes the human genome specifically human, since it offers some of the clearest available evidence that changes to regulatory sequence, rather than changes to protein-coding genes themselves, may have driven a meaningful share of human evolutionary divergence from other primates.

In 2006, two independent research teams, one led by Katherine Pollard and David Haussler at the University of California, Santa Cruz, and another led by Shyam Prabhakar and Edward Rubin at Lawrence Berkeley National Laboratory, published genome-wide searches for a specific and unusual category of DNA sequence: regions that are highly conserved across the genomes of many different vertebrate species, meaning they have remained nearly unchanged over hundreds of millions of years of evolution and are therefore presumed to be functionally important, but that show a surprising, statistically significant burst of sequence change specifically within the human lineage since humans diverged from chimpanzees. Researchers named these regions Human Accelerated Regions, commonly abbreviated HARs, and their basic logic is worth spelling out because it inverts the usual evolutionary conservation argument this book has relied on elsewhere. A sequence conserved across nearly all vertebrates almost certainly does something functionally important, since otherwise it would have drifted apart between species over such an enormous span of evolutionary time. A conserved sequence that then changed rapidly and specifically along the single lineage leading to modern humans is therefore a strong candidate for a sequence whose function was specifically modified during human evolution, potentially contributing to a trait that distinguishes humans from other primates.

The results of these searches were striking in one specific respect: the overwhelming majority of identified HARs, by most published estimates surrounding ninety six percent, fall outside protein-coding sequence entirely, landing instead within non-coding regions, and a substantial proportion show hallmarks of developmental enhancer activity specifically. This finding lent considerable empirical weight to a hypothesis that had been proposed on more theoretical grounds decades earlier, most notably by the geneticists Mary-Claire King and Allan Wilson in an influential 1975 paper, who argued that the striking similarity between human and chimpanzee protein-coding sequence, subsequently confirmed to exceed ninety eight percent identity at the protein level, was difficult to reconcile with the substantial anatomical and behavioral differences between the two species, and that regulatory changes, altering when and where existing genes are switched on rather than changing the genes themselves, were more likely to account for the bulk of human-specific evolutionary change. HARs offered, for the first time, a concrete, genome-wide catalogue of specific candidate regulatory sequences where exactly this kind of regulatory evolution appeared to have occurred.

Individual HARs have since been functionally characterized in enough mechanistic detail to move well beyond the original correlational discovery. The single most extensively studied example, designated HACNS1 and located near a gene called GBX2, was shown through direct transgenic reporter experiments, in which researchers attach a candidate enhancer sequence to a reporter gene and observe where in a developing embryo that reporter becomes active, to function as a limb development enhancer with a pattern of activity specifically altered in the human version compared to the chimpanzee and rhesus macaque versions of the same sequence. Follow-up research published in 2021 and 2022, using a genetically humanized mouse model in which researchers directly replaced the mouse's own version of this enhancer with the human HACNS1 sequence, confirmed that the human variant alters expression of GBX2 during limb development in living tissue and increased activity specifically within the chondrogenic mesenchyme, the developing tissue responsible for forming skeletal cartilage, offering direct experimental evidence connecting a specific, dated human regulatory mutation to a specific, measurable developmental consequence in the limb, a level of mechanistic proof rarely available for hypotheses about human evolutionary divergence.

A second particularly well-studied HAR, designated HAR1, turned out not to function as a conventional enhancer at all, but as a non-coding RNA gene, expressed specifically within a population of neurons active during a critical early window of human cortical development, adding this sequence to the broader catalogue of functionally important long non-coding RNA genes covered in a later chapter of this book. Because HAR1's expression pattern coincides with a developmental period important for the layered organization of the human neocortex, researchers have proposed that this particular accelerated region may have played a role in the evolution of human-specific brain architecture, though, in keeping with this book's commitment to representing genuine uncertainty honestly, the precise functional consequences of HAR1's human-specific sequence changes remain considerably less fully worked out experimentally than those of HACNS1, and this connection to brain evolution specifically should be understood as a well-motivated, actively investigated hypothesis rather than a fully settled finding.

Taken together, promoters, enhancers, silencers, and the three-dimensional chromatin architecture that brings them into functional contact with their target genes constitute what many genome biologists now consider the primary regulatory layer responsible for the staggering diversity of cell types your single genome is capable of producing. It is worth returning, briefly, to the puzzle that opened this chapter: how does one genome build both a neuron and a skin cell. The regulatory elements described in this chapter are the substantial majority of the answer. Different combinations of enhancers become active in different cell types, driven by different combinations of transcription factor proteins present in each specific cellular context, activating different subsets of genes from the identical underlying genome, while different silencers actively suppress genes inappropriate to that particular cell type. A neuron and a skin cell do not differ because they carry different genetic instructions. They differ because an elaborate, largely non-coding regulatory system reads the identical instruction set in dramatically different ways, and the overwhelming majority of the DNA sequence responsible for making that selective reading possible lies entirely outside the protein-coding genes that dominate how most people still picture their own genome.

This regulatory landscape also happens to be where an enormous fraction of disease-associated genetic variation identified through modern genomic research turns out to be located, a point this book raised briefly in its opening chapter and will return to with considerably more specificity in a chapter examining how genome-wide association studies have reshaped our understanding of common, complex disease risk. For now, the essential takeaway is this: the genes themselves, the protein-coding one percent this book's title implicitly sets in contrast to everything else, are, in a meaningful sense, simply parts. The instructions for when, where, and how intensely to use those parts, the actual decision-making layer of your genetic code, live almost entirely within the regulatory shadows this chapter has begun to illuminate.

CHAPTER Nine

RNA That Never Becomes Protein

A cell can build an RNA molecule more than seventeen thousand letters long, invest real metabolic energy transcribing it, and never intend for that molecule to become a protein at all. For most of the twentieth century, this idea would have struck molecular biologists as close to nonsensical. RNA existed, in the standard understanding, as a functional intermediate: DNA gets transcribed into messenger RNA, messenger RNA gets translated into protein, and protein does the actual biological work. RNA molecules that were never translated seemed, by this logic, to be either mistakes or curiosities rather than a genuinely important category of functional molecule in their own right.

That assumption has been thoroughly overturned, and this chapter tells the story through two of the most extensively studied examples in the entire field: XIST, discovered in the early 1990s, and HOTAIR, discovered considerably later, in 2007. Both are long non-coding RNAs, commonly abbreviated lncRNAs, a broad category defined simply as RNA transcripts longer than two hundred nucleotides that do not code for protein. Both illustrate, through very different mechanisms, how thoroughly wrong the old assumption turned out to be.

Every typical human female cell contains two X chromosomes, compared to the single X and single Y found in typical male cells. This creates a potential dosage problem, because genes located on the X chromosome would, without some correcting mechanism, be expressed at twice the level in female cells compared to male cells, a mismatch that would likely cause serious developmental problems given how sensitive many biological processes are to precise gene dosage. Mammalian biology solves this problem through a process called X chromosome inactivation, in which each cell in a female embryo, early in development, essentially and almost entirely at random, silences one of its two X chromosomes, compressing it into a densely packed, largely inactive structure for the remainder of that cell's lifetime and the lifetime of all its descendant cells. The result is that female mammals are genetic mosaics, with roughly half of their cells expressing genes from the X chromosome inherited from their mother and roughly half expressing genes from the X chromosome inherited from their father, an arrangement with documented consequences for a range of biological traits, including, notably, the coat coloring patterns of calico and tortoiseshell cats, whose distinctive patchy coloring is a direct, visible consequence of random X inactivation affecting the pigmentation genes located on their X chromosomes.

The master switch controlling this entire process is not a protein-coding gene. It is XIST, short for X-inactive specific transcript, a long non-coding RNA first identified in the early 1990s through the work of several research groups, notably including work led by Huntington Willard and separately by Neil Brockdorff and colleagues in the United Kingdom. XIST is transcribed exclusively from the specific X chromosome that a cell has selected for inactivation, and the RNA molecule it produces does something that, at the time of its discovery, had no clear precedent in molecular biology. Rather than diffusing away from its site of transcription the way most RNA molecules do, XIST RNA remains physically associated with the chromosome that produced it, gradually spreading outward to coat the entire chromosome, functioning something like a molecular blanket draped specifically and exclusively across one of the cell's two X chromosomes.

Once XIST RNA has coated the target chromosome, it recruits a cascade of additional proteins responsible for compacting chromatin into a transcriptionally silent state, including the Polycomb repressive complex, which deposits a specific chemical modification on the histone proteins around which DNA is wound, a modification strongly associated with stable, long-term gene silencing. Through this recruitment process, XIST converts an entire chromosome, containing well over a thousand protein-coding genes, from an actively transcribed state into a permanently condensed, silenced structure visible under a microscope as a distinct, compact body within the nucleus, historically known to cytologists, well before its molecular mechanism was understood, as the Barr body, after the Canadian scientist Murray Barr, who first described this structure in 1949, more than four decades before XIST itself was identified as the molecule responsible for creating it.

XIST's discovery mattered enormously beyond its specific role in X chromosome biology, because it provided the first clear, mechanistically detailed proof that a non-coding RNA molecule could serve as the master regulatory switch controlling a fundamental developmental process affecting an entire chromosome, using a mechanism, direct physical coating combined with recruitment of chromatin-modifying protein complexes, that no protein-coding gene could plausibly replicate on its own. It opened the conceptual door to a broader question that would occupy genome biologists for the following decades: how many other long non-coding RNAs exist, performing comparably important regulatory functions through comparably novel mechanisms, that had simply gone unnoticed because researchers were not yet looking for this category of molecule?

HOTAIR, discovered in 2007 by researchers including John Rinn, working in Howard Chang's laboratory at Stanford University, provided an early and particularly striking answer to that question, and it demonstrated a mechanism even stranger than XIST's, because HOTAIR regulates genes located on an entirely different chromosome from the one that produces it, something biologists describe as acting in trans rather than in cis.

HOTAIR is transcribed from within the HOXC gene cluster, located on human chromosome twelve. The HOX gene clusters are a well-studied and evolutionarily ancient family of protein-coding genes responsible for controlling body plan development across an enormous range of animal species, determining, broadly speaking, which structures develop in which positions along an organism's head-to-tail axis during embryonic development. Given HOTAIR's location within this developmentally critical gene cluster, researchers initially expected it might regulate nearby genes within the same HOXC cluster on chromosome twelve. What Rinn and colleagues found instead surprised the field considerably: HOTAIR RNA, once produced, travels away from its site of transcription and represses an entirely separate cluster of HOX genes, the HOXD cluster, located on a completely different chromosome, chromosome two.

The mechanism, worked out through subsequent research over the following several years, involves HOTAIR functioning as a molecular scaffold, a physical platform bringing together two separate protein complexes that would not otherwise efficiently interact. One end of the HOTAIR RNA molecule binds Polycomb repressive complex 2, the same chromatin-silencing complex recruited by XIST in the X inactivation mechanism described above. The other end of the HOTAIR molecule binds a distinct protein complex called LSD1, which removes a different chemical modification from histone proteins, one normally associated with active gene expression. By physically holding both complexes together in a single molecular assembly and guiding that assembly to specific locations within the HOXD cluster, HOTAIR effectively delivers a coordinated, two-pronged silencing signal to its distant target genes, a level of molecular choreography that a diffusing RNA molecule, rather than a protein, appears uniquely well suited to accomplish, given RNA's capacity to fold into complex three-dimensional shapes offering multiple distinct binding surfaces along a single molecule.

HOTAIR's biological significance extends well beyond basic developmental biology, because subsequent research through the 2010s repeatedly identified elevated HOTAIR expression as a documented marker of poor prognosis across numerous human cancer types, including breast cancer, colorectal cancer, and hepatocellular carcinoma, among others, with elevated HOTAIR levels specifically associated with increased tumor invasiveness and metastatic spread in multiple independent clinical studies. The proposed mechanism connecting HOTAIR to cancer progression builds directly on its normal developmental function: by inappropriately reprogramming chromatin states at genes well beyond its usual developmental targets, dysregulated HOTAIR expression appears capable of activating gene expression programs that favor cancer cell invasion and spread into surrounding tissue, essentially hijacking the same molecular scaffold mechanism responsible for normal HOX gene regulation and redirecting it toward genes relevant to cancer progression instead. This research has generated genuine, ongoing clinical interest in HOTAIR as both a prognostic biomarker, helping predict how aggressively a given patient's cancer is likely to behave, and as a potential therapeutic target, with several research groups actively investigating whether directly blocking HOTAIR activity might slow cancer progression in specific tumor types where its expression is characteristically elevated.

XIST and HOTAIR represent two of the most thoroughly characterized examples within what has become, since HOTAIR's 2007 discovery in particular, a rapidly expanding catalogue of functionally significant long non-coding RNAs. Current estimates place the total number of distinct long non-coding RNA genes in the human genome somewhere in the range of fifteen to twenty thousand, a figure comparable to, and by some counting methods exceeding, the total number of protein-coding genes, though it should be said plainly that the functional roles of the substantial majority of these individual long non-coding RNAs remain incompletely characterized, and ongoing research continues actively working through this backlog, gene by gene, mechanism by mechanism, using experimental approaches this book will examine in more detail in a later chapter devoted specifically to the newest generation of functional genomics tools.

What XIST and HOTAIR together demonstrate, with a level of mechanistic detail that leaves little room for reasonable scientific dispute, is that RNA molecules never intended to become protein can nonetheless perform essential, specific, and in some cases medically consequential biological functions, using molecular strategies, direct physical chromosome coating, long-distance scaffold-mediated protein recruitment, that protein-coding genes are simply not structurally equipped to replicate. The category of molecule that twentieth century molecular biology largely dismissed as an intermediate step on the way to something more important has turned out, in a substantial number of specific, well-documented instances, to be the important thing itself.

CHAPTER Ten

The Small Molecules With Big Reach

In 1993, two researchers studying an unglamorous, millimeter-long roundworm published a finding so far outside the mainstream concerns of molecular biology at the time that it took the field roughly seven years to recognize its full significance. The researchers were Victor Ambros and Gary Ruvkun, the organism was Caenorhabditis elegans, and the discovery would eventually earn both scientists the 2024 Nobel Prize in Physiology or Medicine, awarded jointly for the discovery of microRNA and its role in gene regulation, more than three decades after their initial 1993 publication. A related but mechanistically distinct discovery, RNA interference, made in the same worm species using longer double stranded RNA molecules, had already earned a separate Nobel Prize for Andrew Fire and Craig Mello in 2006, and the two discoveries are often confused with one another despite operating through different molecular pathways.

The story begins with a mutant strain of C. elegans that displayed an unusual developmental timing defect, progressing through its larval developmental stages incorrectly. Ambros, working to identify the genetic cause of this defect, traced the problem to a gene called lin-4. What he found when he examined lin-4 closely was genuinely puzzling given the standard understanding of gene function at the time. Lin-4 did not encode a protein. It produced only a very short RNA molecule, roughly twenty two nucleotides long, far too small to plausibly function as a messenger RNA capable of directing protein synthesis through the normal translation machinery.

Working in a separate laboratory but investigating an interconnected piece of the same developmental pathway, Ruvkun had been studying a different gene, lin-14, whose protein product needed to decrease at a specific, precisely timed point during the worm's larval development in order for normal developmental progression to occur. Comparing notes, Ambros and Ruvkun realized that the short RNA sequence produced by lin-4 was partially complementary to a specific region within the messenger RNA produced by lin-14, meaning the two RNA sequences could, based on the standard base-pairing rules governing how RNA molecules interact, physically bind directly to one another.

This observation, published jointly by the two laboratories in the same December 1993 issue of the journal Cell, described an entirely new mechanism of gene regulation. The short lin-4 RNA molecule was binding directly to specific sites within the lin-14 messenger RNA, and this binding somehow prevented the lin-14 message from being translated into protein, without destroying the messenger RNA molecule itself or altering the underlying lin-14 gene sequence in any way. A small, non-coding RNA molecule was directly controlling the protein output of a specific target gene, through simple, sequence-based molecular recognition, at a scale and through a mechanism nobody in the field had previously documented.

For several years following the 1993 publication, this finding was largely treated as an unusual, perhaps idiosyncratic feature specific to C. elegans developmental biology, rather than as the first documented example of a broadly conserved, fundamentally important regulatory mechanism operating across the animal kingdom. That perception changed decisively in 2000, when Ruvkun's laboratory identified a second small regulatory RNA, called let-7, performing an analogous regulatory function at a later stage of C. elegans development, and, critically, found that the let-7 sequence itself was conserved not just in nematode worms but across an enormous range of animal species, including flies, mollusks, and, significantly, humans. This conservation across such evolutionarily distant species strongly indicated that the mechanism Ambros and Ruvkun had first identified in a single obscure worm gene was not a developmental curiosity confined to nematodes, but an ancient, fundamentally important regulatory system present throughout the animal kingdom, including in human cells.

The subsequent years, particularly the early 2000s, saw an explosion of research identifying additional small regulatory RNAs of this general category, now formally named microRNAs, across a rapidly expanding range of species and biological contexts. Current, well-established estimates place the total number of distinct microRNA genes encoded in the human genome at somewhere over two thousand, and computational and experimental studies suggest that microRNAs, collectively, regulate the expression of a substantial majority of all human protein-coding genes, with widely cited estimates suggesting more than sixty percent of human genes are subject to microRNA-mediated regulation through at least one binding site conserved across mammalian evolution.

The molecular mechanism, worked out in considerably more biochemical detail over the years following the original 1993 and 2000 discoveries, involves a multi-step processing pathway. MicroRNA genes are initially transcribed as considerably longer RNA precursor molecules, which fold back on themselves to form a characteristic hairpin-shaped structure due to internal sequence complementarity. This precursor structure is progressively trimmed down by two specialized enzymes, first an enzyme called Drosha, operating within the cell nucleus, and subsequently an enzyme called Dicer, operating in the cell's cytoplasm, ultimately yielding the short, mature, roughly twenty-two-nucleotide microRNA sequence capable of directing gene regulation. This mature microRNA is loaded into a larger protein assembly called the RNA-induced silencing complex, commonly abbreviated RISC, which uses the microRNA sequence as a guide to locate complementary target sequences within messenger RNA molecules throughout the cell. Once RISC, guided by its loaded microRNA, locates and binds a matching target sequence, it can trigger either direct destruction of the target messenger RNA molecule or, alternatively, simply block that molecule from being translated into protein, depending on the precise degree of sequence complementarity between the microRNA and its target site, with animal microRNAs, including human ones, more commonly employing the translational blocking mechanism through imperfect rather than perfect sequence matches.

A single microRNA can target and regulate dozens, sometimes hundreds, of distinct messenger RNA molecules, because the sequence matching required for functional targeting does not need to be perfect across the microRNA's full length, only sufficiently precise within a specific short stretch called the seed region. This one-to-many targeting relationship allows a relatively modest total number of microRNA genes, just over two thousand in the human genome, to collectively exert regulatory influence across the substantial majority of the genome's protein-coding output, functioning as a genome-wide fine-tuning system capable of adjusting gene expression levels with considerably finer-grained precision than the simpler on-or-off regulatory logic of promoters and enhancers alone could achieve.

The medical significance of microRNA biology has grown substantially since the initial 1993 and 2000 discoveries, moving well beyond basic developmental biology into direct clinical relevance across a range of human diseases.

Specific microRNAs have been repeatedly documented as dysregulated, meaning produced at abnormally high or low levels, across numerous cancer types, with some microRNAs functioning as documented tumor suppressors when their normal expression is lost, and others functioning as documented oncogenes when their expression becomes abnormally elevated. This has generated substantial ongoing research interest in using microRNA expression patterns as diagnostic and prognostic biomarkers across multiple cancer types, and separately, in developing microRNA-targeting therapeutics designed to restore normal regulatory function in diseases where specific microRNA dysregulation has been clearly documented as a contributing factor.

One particularly well-developed example involves miR-122, a microRNA expressed at high levels specifically in liver tissue, which the hepatitis C virus was found to depend on directly for efficient replication within infected liver cells, since the virus's own RNA genome contains a binding site for miR-122 that appears necessary for stable viral RNA accumulation within host cells. This specific, well-documented dependency led to the clinical development of miravirsen, an experimental drug designed to block miR-122 activity and thereby interfere with hepatitis C viral replication, representing one of the more advanced examples of a therapeutic approach built directly on basic microRNA biology first described in that 1993 C. elegans paper. Miravirsen successfully reduced viral RNA levels in phase two human clinical trials published in 2013, offering direct clinical proof that a microRNA-targeting drug could work in humans, though the drug's development was later discontinued in favor of newer antiviral approaches for hepatitis C that had by then become available.

It is worth pausing, at the close of this chapter, to register how improbable this entire scientific trajectory would have seemed to most molecular biologists working before 1993. A twenty-two-letter RNA fragment, produced by a gene in a microscopic worm, too small to encode any conceivable protein, turned out to represent the founding example of a regulatory mechanism now understood to influence the expression of a majority of human genes, implicated in cancer biology across dozens of distinct tumor types, and actively pursued as a therapeutic target against a major human viral disease. Ambros and Ruvkun did not set out, in the early 1990s, to discover a fundamental new layer of gene regulation operating throughout the animal kingdom. They set out to understand why a particular strain of roundworm developed incorrectly. The gap between that modest original question and the eventual scale of what it revealed stands as one of the clearer illustrations, among many this book has offered so far, of how much biological significance had been hiding, entirely unsuspected, within the non-coding regions of the genome that earlier generations of scientists had largely agreed to stop looking at closely.

CHAPTER Eleven

Pseudogenes, the Genome's Fossils

Somewhere in your genome sits a gene that looks, on paper, almost identical to one of the most important tumor-suppressing genes your cells rely on. It has roughly the same sequence. It occupies a plausible-looking genomic location. And for decades, it was classified as biologically dead on arrival, a broken copy incapable of producing anything useful. That classification, as this chapter will show, was wrong in at least one well-documented and medically important case, and the story of how researchers discovered the mistake reshapes how we should think about an entire category of genomic sequence: pseudogenes.

A pseudogene, in the most basic definition, is a DNA sequence that closely resembles a known functional gene but that appears, through one specific defect or another, incapable of producing a normal, functional protein. Pseudogenes arise through a small number of well-understood molecular mechanisms, and understanding these mechanisms matters because different origins tend to produce pseudogenes with different downstream properties.

The first major category, called unprocessed or duplicated pseudogenes, arises when a functional gene gets accidentally duplicated somewhere in the genome, and one of the resulting two copies subsequently accumulates disabling mutations over evolutionary time, degrading into a nonfunctional pseudogene while its sibling copy continues performing the original gene's job. Because gene duplication is a common evolutionary event, and because a duplicated gene copy is initially redundant and therefore under relaxed selective pressure to remain intact, this pathway has generated a substantial number of pseudogenes scattered throughout the human genome, each retaining a family resemblance to whichever functional gene it originally duplicated from.

The second major category, called processed pseudogenes, arises through an entirely different mechanism, one that connects directly back to the transposable elements covered earlier in this book. Recall that LINE-1 elements encode a reverse transcriptase enzyme capable of converting RNA back into DNA and inserting that DNA copy elsewhere in the genome. This same LINE-1 machinery occasionally, and evidently accidentally, acts on messenger RNA transcripts from entirely unrelated, functional protein-coding genes, reverse-transcribing them into DNA and inserting that DNA copy into a new genomic location. Because this process starts from a mature, already-processed messenger RNA molecule, the resulting pseudogene characteristically lacks introns, even though its functional parent gene typically retains them, and it typically lacks a proper promoter sequence at its new insertion site, since it was never designed by evolution to be transcribed as an independent gene in its own right. Processed pseudogenes generated through this mechanism are consequently often assumed to be transcriptionally silent from the moment of their creation, inserted essentially at random into whatever genomic neighborhood the LINE-1 insertion machinery happened to select, with no functional promoter to drive their own expression.

Current estimates place the total number of pseudogenes in the human genome at somewhere around fourteen thousand, a substantial number given that it approaches, and by some counting conventions exceeds, the total number of functional protein-coding genes. For decades, the default assumption regarding this entire category, consistent with the broader junk DNA framework this book has traced from its origins, was straightforward: pseudogenes are broken genes, evolutionary leftovers, genuinely and definitively nonfunctional, existing as molecular fossils documenting past evolutionary events but contributing nothing to the current biology of the organism carrying them.

This assumption held up reasonably well for most pseudogenes examined in detail, and it is important to state plainly that this remains the accurate characterization for the substantial majority of documented pseudogenes. Most pseudogenes, when carefully studied, do appear to be transcriptionally silent or functionally inert, consistent with the straightforward assumption that they are simply broken, historical remnants of gene duplication or LINE-1-mediated retrotransposition events. This chapter is not arguing that pseudogenes are secretly all functional; that would overstate the evidence just as badly as the original blanket dismissal did. What this chapter argues, instead, is that a meaningful and medically significant minority of pseudogenes have been shown, through rigorous, specific experimental evidence, to perform genuine regulatory functions, and that this minority is important enough, and well-documented enough, to warrant serious attention.

The most extensively studied and medically significant example involves PTEN, one of the most important tumor suppressor genes in the human genome, whose normal function involves restraining cell growth and division, and whose loss or inactivation is documented as a contributing factor across a wide range of human cancers. PTEN has a well-characterized pseudogene, designated PTENP1, arising through the duplication mechanism described above, sharing extensive sequence similarity with the functional PTEN gene but containing specific mutations that prevent it from producing a functional PTEN protein.

Research published in a landmark 2010 paper in the journal Nature, led by Pier Paolo Pandolfi's laboratory, demonstrated that PTENP1, despite its inability to produce functional protein, is nonetheless actively transcribed into RNA in human cells, and that this pseudogene-derived RNA performs a genuinely important regulatory function protecting the expression of its functional PTEN counterpart. The mechanism involves competition for microRNA binding, connecting directly back to the regulatory system explored in the previous chapter. Several microRNAs that normally bind to and suppress PTEN messenger RNA can also bind to the highly similar PTENP1 pseudogene transcript, because the two sequences remain similar enough for the same microRNAs to recognize both. By absorbing a share of this microRNA-mediated suppression that would otherwise be directed entirely at the functional PTEN transcript, PTENP1 effectively acts as what researchers termed a molecular decoy or sponge, soaking up microRNA activity and thereby helping protect the functional PTEN gene's expression from being suppressed as heavily as it otherwise would be.

The medical significance of this mechanism became directly apparent when the same research demonstrated that PTENP1 expression is reduced or lost in specific human cancers, and that this loss correlates with correspondingly reduced PTEN protein expression, consistent with the pseudogene's protective, decoy function being genuinely important for maintaining adequate levels of this critical tumor suppressor in normal, non-cancerous tissue. A DNA sequence long classified as a broken, nonfunctional copy of a cancer-fighting gene turned out to be actively helping fight cancer itself, through a regulatory mechanism nobody had anticipated before the underlying biology of microRNA sponging was worked out.

The PTENP1 discovery was not an isolated curiosity. Subsequent research through the 2010s identified a growing list of additional pseudogenes performing documented regulatory functions through broadly comparable mechanisms, competing for microRNA binding, or, in a related but mechanistically distinct pathway, producing small interfering RNA molecules capable of directly suppressing their functional parent gene through a different regulatory pathway entirely. A pseudogene derived from the gene encoding a protein called NANOG, important in maintaining stem cell identity, and pseudogenes associated with several other developmentally important genes, have similarly been documented performing regulatory functions relevant to their parent gene's expression, though the specific mechanisms vary somewhat from case to case and remain under active investigation in several instances.

It is worth being explicit, in closing this chapter, about exactly how much this reframes the broader pseudogene category and how much it does not. It would be a significant overstatement to claim that pseudogenes, as a class, have been shown to be broadly functional; the specific, well-documented examples covered in this chapter represent a genuine minority carved out of a much larger category where the original assumption of nonfunctionality continues to hold up reasonably well under scrutiny. What has changed, and changed in a way with real clinical stakes, is the default assumption a careful researcher should now bring to any individual pseudogene before dismissing it. The PTENP1 story specifically demonstrates that a sequence's inability to produce a working protein does not automatically mean that sequence, transcribed as RNA rather than translated into protein, cannot still perform a biologically and medically important function, in this specific documented case, a function directly relevant to how effectively your cells suppress cancer development. The genome's fossils, it turns out, are not all equally dead. Some of them, examined closely enough, are still quietly doing work.

CHAPTER Twelve

Telomeres and Centromeres

Every time one of your cells divides, it faces a peculiar structural problem that has nothing to do with genes in the conventional sense, and everything to do with the basic chemistry of how DNA gets copied. The enzyme responsible for DNA replication, DNA polymerase, cannot fully replicate the very end of a linear DNA molecule. Each time a chromosome is copied, a small amount of sequence at its extreme tip goes uncopied and is effectively lost. Left unaddressed, this problem would mean that every cell division shortens your chromosomes slightly, generation after generation, cell division after cell division, until essential genes near the chromosome ends eventually began to be damaged or lost entirely.

Evolution's solution to this problem, present across nearly all eukaryotic organisms, involves capping the ends of every chromosome with a specialized, highly repetitive, non-coding DNA sequence called a telomere. In humans, the telomeric sequence consists of the six-letter unit TTAGGG, repeated over and over, thousands of times in succession, at the very tip of every chromosome. This repetitive buffer sequence does not encode any protein and does not, by the standard definitions applied to protein-coding genes, do anything a gene does. Its function is instead purely structural and protective, providing an expendable buffer of sequence that can be gradually shortened through the normal replication process described above without any loss of genuinely important genetic information, since the actual protein-coding genes sit safely further inward, well clear of the chromosome tip where the shortening occurs.

The discovery of how telomeres are actually maintained, rather than simply shortening indefinitely with every cell division until a chromosome's protective buffer runs out entirely, involved one of the more elegant experimental research programs in modern molecular biology, and it earned its principal investigators the 2009 Nobel Prize in Physiology or Medicine, awarded jointly to Elizabeth Blackburn, Carol Greider, and Jack Szostak. Blackburn, working in the late 1970s and early 1980s on the unusual single-celled pond organism Tetrahymena, first characterized the specific repetitive sequence structure of telomeric DNA. Working subsequently with Szostak, whose parallel research on yeast chromosome stability provided a complementary experimental system, the two researchers demonstrated that telomeric sequence from one organism, Tetrahymena, could functionally protect chromosome ends when transplanted into an entirely different organism, yeast, strong evidence that telomeres represented a conserved, actively maintained biological mechanism rather than an incidental structural quirk specific to any single species.

The mechanism maintaining telomere length was worked out through subsequent research led by Carol Greider, then a graduate student in Blackburn's laboratory, who in 1984 identified an enzyme, now called telomerase, capable of actively adding new copies of the repetitive telomeric sequence onto the ends of chromosomes, directly counteracting the gradual shortening that would otherwise occur through ordinary DNA replication. Telomerase accomplishes this through a genuinely unusual mechanism: it carries its own internal RNA template, a short non-coding RNA molecule built directly into the enzyme's structure, which it uses as a guide for synthesizing new telomeric DNA sequence, essentially functioning as a specialized reverse transcriptase dedicated specifically to chromosome end maintenance.

Telomerase activity is not uniform across all the cells in your body, and this variation turns out to carry significant implications for both aging and cancer, the two major biological themes this chapter will connect telomere biology to. In most normal, differentiated adult human cells, telomerase activity is low or entirely absent, meaning telomeres in these cells do progressively shorten with each successive cell division, functioning as a kind of internal, cumulative record of how many times a given cell lineage has divided across a person's lifetime. Once a cell's telomeres shorten below a critical threshold length, the cell typically enters a state called replicative senescence, permanently ceasing further division, a phenomenon first documented experimentally by Leonard Hayflick in the early 1960s, well before telomeres or telomerase had been identified as the underlying mechanism, and still commonly referred to as the Hayflick limit in his honor.

This connection between telomere shortening and cellular aging has generated substantial research interest in telomere length as a potential biomarker of biological aging, distinct from simple chronological age, and a considerable body of published epidemiological research has documented associations between shorter average telomere length, measured in blood cell samples, and increased risk for various age-associated conditions, including cardiovascular disease. It is important, in the interest of scientific accuracy, to note that this remains an area of active research with some genuinely unresolved questions regarding causation versus correlation. Whether telomere shortening is itself a primary driver of age-related decline, or whether it functions more as a downstream marker reflecting other underlying aging processes, or some combination of both operating simultaneously, remains a subject of legitimate ongoing scientific investigation rather than a fully settled question, and readers should treat any confident, simplified claims connecting telomere length directly to aging as somewhat oversimplified relative to the genuine complexity documented in the peer-reviewed literature on this subject.

The relationship between telomerase and cancer presents an equally important, and in some ways inverse, story. While most normal adult cells keep telomerase activity switched off, allowing telomeres to progressively shorten as a natural check against unlimited cell division, the substantial majority of human cancers, by most published estimates somewhere around eighty five to ninety percent of tumors examined, have been documented to reactivate telomerase activity, allowing cancer cells to bypass the normal telomere-shortening limit on cell division entirely and divide indefinitely, a hallmark capability that cancer biologists formally recognize as one of the defining features distinguishing malignant cancer cells from normal, appropriately constrained tissue. This documented connection between telomerase reactivation and cancer's capacity for unlimited division has made telomerase a longstanding and still actively pursued target for cancer drug development, with multiple therapeutic approaches designed to inhibit telomerase activity specifically in cancer cells having progressed through various stages of clinical research.

Centromeres, the second major category of repetitive structural DNA covered in this chapter, serve an entirely different but equally essential purpose, located not at the ends of chromosomes but at a specific internal position, typically visible under a microscope as the pinched, narrowed region where the two halves of a duplicated chromosome remain joined together during cell division. The centromere's essential function is providing the physical attachment point where a large protein structure called the kinetochore assembles, and it is this kinetochore structure that ultimately connects the chromosome to the cellular machinery responsible for physically pulling duplicated chromosome copies apart into the two resulting daughter cells during cell division, ensuring that each new cell receives a complete and accurate set of chromosomes.

Human centromeric DNA consists overwhelmingly of a specific repetitive sequence family called alpha satellite DNA, built from a repeating unit roughly one hundred seventy one base pairs long, itself repeated in tandem, one copy after another, thousands of times in succession, producing centromeric regions that can extend across millions of base pairs of largely repetitive, non-coding sequence. For decades, this highly repetitive centromeric sequence proved genuinely difficult to sequence accurately using the technology available at the time, since standard genome sequencing methods struggle considerably with long stretches of near-identical repeated sequence, making it difficult to determine the precise order and exact number of repeat units present. This technical limitation meant that even after the original Human Genome Project was declared essentially complete in 2003, and even after a subsequent "finished" reference genome was published in 2004, centromeric regions across all human chromosomes remained substantially unresolved gaps in the official reference sequence, a limitation that persisted for nearly two additional decades.

This gap was finally closed through the Telomere-to-Telomere consortium, a large collaborative research effort that published the first genuinely complete, gapless sequence of an entire human genome, including full centromeric and telomeric regions, in a set of papers in the journal Science in 2022, made possible by newer long-read sequencing technologies capable of reading through repetitive stretches that had defeated earlier sequencing methods. This achievement, arriving more than two decades after the original Human Genome Project's celebrated announcement, filled in roughly the last eight percent of the human genome that had remained unresolved, and the overwhelming majority of that final, hardest-to-sequence eight percent consisted precisely of the repetitive centromeric and other structurally complex sequence this chapter has been describing, a fitting testament to just how technically demanding this category of non-coding DNA has proven to study properly, even as its basic biological importance, holding chromosomes together correctly during every single cell division your body performs, was never seriously in doubt.

Telomeres and centromeres, taken together, illustrate a category of non-coding function distinct from the regulatory and RNA-based mechanisms covered in earlier chapters of this book. Neither structure works by producing a functional RNA molecule, and neither works by regulating the expression of some other gene. Their function is purely structural and mechanical: providing an expendable protective buffer at chromosome ends, and providing a stable, reliable attachment point for the cellular machinery that physically separates chromosomes during division. Both functions depend entirely on repetitive, highly conserved, but genuinely non-coding sequence, and both illustrate, from yet another angle, how thoroughly the operational integrity of your genome, its ability to be copied accurately and divided correctly across billions of cell divisions occurring inside your body across your entire lifetime, depends on precisely the category of sequence early genetics assumed did not much matter.

CHAPTER Thirteen

When the Silence Breaks

A geneticist reading a patient's DNA sequence for Huntington's disease is not looking for a typo. Most genetic diseases, the kind covered in introductory biology courses, arise from a single altered letter, or a handful of altered letters, disrupting a gene's protein-coding sequence in a way that changes or destroys the resulting protein's structure and function. Huntington's disease works differently, and understanding why requires returning to a specific, strange category of DNA sequence that sits at the center of this chapter: short sequences of DNA, sometimes just three letters long, repeated over and over in direct succession, where the exact number of repetitions, rather than the sequence itself, is what determines whether a person develops a devastating, currently incurable neurological disease.

The gene responsible for Huntington's disease, called HTT, contains a stretch of DNA where the three-letter sequence CAG repeats in direct tandem succession. In people without Huntington's disease, this CAG repeat typically occurs somewhere between ten and thirty five times in a row, a range considered normal and stable across generations. When the repeat count climbs to forty or more, the person will, with essentially complete certainty given a long enough lifespan, develop Huntington's disease, a progressive neurodegenerative condition causing movement abnormalities, cognitive decline, and psychiatric symptoms, typically emerging in mid-adulthood and worsening steadily over one to two subsequent decades. The gene responsible was identified in 1993 through a decade-long collaborative research effort, one of the earlier major successes of what was then a young and still developing field of human genetic disease mapping.

What makes this mechanism scientifically distinctive, and directly relevant to this book's broader argument, is that the CAG repeat responsible for Huntington's disease does actually fall within the HTT gene's protein-coding sequence, translating directly into a stretch of repeated glutamine amino acids within the resulting huntingtin protein. An expanded repeat produces a protein carrying an abnormally long glutamine stretch, and this elongated protein segment causes the resulting protein to fold incorrectly and clump together inside neurons in a toxic, damaging way. Huntington's disease belongs to a category of conditions called polyglutamine diseases, and while the underlying repeat expansion mechanism connects directly to the broader theme of this chapter, in this particular case the damage occurs through a coding sequence effect rather than a purely non-coding one.

The connection to this book's central subject becomes more direct, and in some ways more striking, when we turn to a second, closely related disease: Fragile X syndrome, the most common inherited cause of intellectual disability worldwide and the most common known single-gene cause of autism spectrum disorder. Fragile X syndrome arises from an expanded CGG repeat located not within the protein-coding sequence of the responsible gene, called FMR1, but within its non-coding regulatory region, specifically in what is called the five prime untranslated region, positioned just upstream of where actual protein-coding sequence begins. In unaffected individuals, this CGG repeat typically occurs somewhere between five and forty four times. When the repeat count expands beyond approximately two hundred repetitions, a threshold researchers call a full mutation, a specific and severe consequence follows: the expanded repeat triggers a chemical modification process called DNA methylation across the surrounding region, and this methylation effectively silences the entire FMR1 gene, shutting off its transcription almost completely, so that essentially no functional FMR1 protein gets produced at all.

This mechanism is worth sitting with, because it represents something genuinely distinct from the Huntington's disease example just described. In Fragile X syndrome, the disease-causing mutation occurs entirely within non-coding sequence, and the mechanism of harm involves not a damaged or misfolded protein, but the complete transcriptional silencing of an otherwise entirely normal, undamaged gene, triggered purely by an expansion in repeat number occurring outside the gene's own coding sequence. The FMR1 protein that would have been produced, had transcription proceeded normally, is itself entirely unremarkable in a full mutation carrier; the problem is that it never gets made in the first place, because the expanded non-coding repeat sequence upstream has triggered a regulatory shutdown of the entire gene.

A third and more recently characterized example extends this pattern even further into genuinely non-coding, intergenic territory, and connects the story directly to the transposable element biology covered earlier in this book. In 2011, researchers identified a massive expansion of a six-letter repeat sequence,

GGGGCC, located within a non-coding region of a gene called C9orf72, as the single most common known genetic cause of both amyotrophic lateral sclerosis, commonly known as ALS or Lou Gehrig's disease, and frontotemporal dementia, accounting for a substantial fraction, in some studied populations approaching forty percent, of cases where these two related neurodegenerative conditions run in families. The repeat expansion responsible sits within an intron of the C9orf72 gene, in a region that, following the logic established in this book's earlier chapter on introns, might once have been assumed to carry little functional consequence given its location outside the gene's actual protein-coding exons.

The mechanisms by which this C9orf72 repeat expansion causes neurodegenerative disease turn out to be unusually varied and are still being actively worked out by researchers, representing a genuinely active area of ongoing neuroscience research rather than a fully resolved textbook mechanism. Current evidence, accumulated across numerous published studies through the 2010s and into the 2020s, points toward at least three distinct, potentially overlapping pathological mechanisms operating simultaneously. First, the expanded repeat sequence, when transcribed into RNA, appears capable of forming abnormal clumps within the cell nucleus, called RNA foci, which sequester and disrupt the normal function of various RNA-binding proteins the cell needs for other essential purposes. Second, and more surprisingly, researchers discovered that the expanded repeat sequence can, through a highly unusual mechanism called repeat-associated non-AUG translation, generate abnormal, toxic protein fragments despite occurring within a region of the gene that is not supposed to be translated into protein at all under normal circumstances, essentially representing translation occurring somewhat outside the standard genetic coding framework the earlier chapters of this book described. Third, the same regulatory disruption that connects back to the intron-splicing biology covered in an earlier chapter appears to interfere with normal processing of the C9orf72 gene's own messenger RNA, potentially reducing production of the normal C9orf72 protein alongside the toxic effects generated by the first two mechanisms.

Stepping back from these three specific, well-documented disease examples, what general lesson does repeat expansion disease offer for this book's broader argument about the non-coding genome? The lesson is that the categories this book has been building, coding versus non-coding, gene versus regulatory element, translated versus untranslated, do not map cleanly onto categories of medical consequence. A disease-causing mutation occurring entirely outside conventional protein-coding sequence, as in Fragile X syndrome and C9orf72-linked ALS, can produce medical consequences every bit as severe, every bit as well-documented, and every bit as clinically significant as a mutation occurring within a classical protein-coding gene. The FMR1 gene in a Fragile X full mutation carrier is not itself damaged or mutated in any conventional sense; it is perfectly intact and would function normally if only it were being transcribed, and the entire pathology traces back to a regulatory failure occurring in sequence that earlier generations of genetics education would have confidently, and wrongly, dismissed as unimportant.

This category of disease has also generated genuine, actively developing therapeutic approaches directly informed by the specific non-coding mechanisms described in this chapter, and one particular example is worth examining closely, because it connects nearly every thread this book has followed so far into a single, remarkable clinical story.

Antisense oligonucleotides, commonly abbreviated ASOs, are short, laboratory-synthesized strands of nucleic acid, typically fewer than twenty five letters long, chemically modified for stability, designed to bind directly to a specific target RNA molecule through the same base-pairing rules that govern how microRNAs recognize their targets, described earlier in this book. Once bound to a disease-relevant messenger RNA, an antisense oligonucleotide can act through several distinct mechanisms depending on its specific chemical design: triggering an enzyme called RNase H to cut and destroy the target RNA, physically blocking the target RNA from being translated, or, in a particularly useful variant, altering how a target RNA gets spliced, allowing researchers to correct certain splicing defects directly by steering the cell's own splicing machinery away from a faulty pattern.

In April 2023, the United States Food and Drug Administration approved tofersen, an antisense oligonucleotide therapy designed to treat a specific inherited form of ALS caused by mutations in a gene called SOD1, which accounts for approximately two percent of all ALS cases. Tofersen works by binding directly to SOD1 messenger RNA and triggering its destruction through the RNase H mechanism described above, thereby reducing production of the toxic, misfolded SOD1 protein responsible for motor neuron damage in this specific genetic form of the disease. Clinical trial results published in the New England Journal of Medicine showed that tofersen substantially reduced both SOD1 protein levels in cerebrospinal fluid and neurofilament light chain levels in blood plasma, a biomarker of ongoing neuronal damage, though the drug's effect on direct clinical measures of disease progression within the twenty eight week trial period did not reach statistical significance, a result the research and patient communities have treated with appropriate caution rather than either dismissal or overstatement. Subsequent longer-term, open-label follow-up data have suggested that patients who began treatment earlier in their disease course experienced slower subsequent decline than those who began later, findings that remain under continued investigation but that have supported tofersen's continued clinical use and ongoing evaluation.

A second, considerably more unusual antisense oligonucleotide story illustrates just how far this technology has progressed toward genuinely individualized medicine, and it connects directly and specifically back to the transposable element biology covered earlier in this book. In 2016, a girl named Mila Makovec, then six years old, was diagnosed with a fatal, rapidly progressing form of Batten disease, a rare neurodegenerative condition causing seizures, vision loss, and progressive loss of motor function. Genetic sequencing of Mila's DNA revealed an unusual cause: one copy of the responsible gene, called MFSD8, carried a conventional disease-causing mutation, but the second copy appeared normal in conventional sequencing and could not initially explain her symptoms. Further investigation revealed that the second copy carried an insertion, within one of the gene's introns, of a SINE-VNTR-Alu element, a composite transposable element belonging to the broader family of retrotransposons this book examined closely in its earlier chapters, and that this specific insertion was disrupting normal splicing of the MFSD8 transcript.

A research team led by Timothy Yu at Boston Children's Hospital designed a custom antisense oligonucleotide, ultimately named milasen after Mila herself, specifically engineered to bind the region of RNA disrupted by her particular transposable element insertion and steer the splicing machinery back toward the normal, functional splicing pattern. Remarkably, the entire process from Mila's diagnosis to receiving her first dose of a drug designed and manufactured specifically and exclusively for her own unique genetic mutation took less than a year, an almost unheard of timeline in conventional drug development, made possible specifically because antisense oligonucleotide chemistry is modular enough that changing the target sequence does not require reinventing the underlying drug platform. Milasen reduced Mila's seizure frequency and stabilized aspects of her condition for a period following treatment, and while Mila's underlying disease had already progressed significantly before treatment began and she passed away in 2021, her case became the founding example cited by the FDA in subsequent draft guidance for developing personalized antisense therapies for other patients with vanishingly rare, individually unique genetic mutations, a regulatory pathway that has since supported additional single-patient antisense drug programs for other rare genetic conditions.

The milasen story draws together nearly every major theme this book has traced across its preceding chapters into a single clinical case: a transposable element insertion, of the same broad category discussed in the chapters on Alu and LINE-1 elements, disrupting intron splicing, the mechanism examined in this book's chapter on introns, treated using antisense RNA technology, built on the same base-pairing biology that governs microRNA function, to correct a non-coding regulatory defect that conventional genetics would have struggled to even properly diagnose only a decade earlier. It stands as one of the clearest available illustrations of how directly the basic, unglamorous science this book has walked through, decades of research into transposable elements, splicing mechanisms, and RNA biology conducted with no specific patient in mind, can translate into an actual treatment, tailored to a single child's specific, individual genome, when the right diagnostic and therapeutic tools finally exist to make that translation possible.

For Huntington's disease, several therapeutic approaches currently in various stages of clinical development aim to reduce production of the toxic expanded huntingtin protein directly, using this same antisense oligonucleotide technology. For Fragile X syndrome, researchers have explored approaches aimed at reactivating the silenced FMR1 gene by reversing the abnormal DNA methylation responsible for shutting it down, an approach that, if eventually successful in clinical application, would directly target and reverse the specific non-coding regulatory mechanism responsible for the disease in the first place, rather than attempting to compensate for its downstream consequences.

The chapters of this book so far have largely told a story of hidden benefit, of non-coding sequence quietly performing essential, often previously unsuspected biological functions. This chapter has been different in tone, deliberately so, because an honest, comprehensive account of the non-coding genome cannot responsibly omit its capacity for serious harm when its normal function breaks down. The same repetitive, non-coding sequence architecture that this book has spent its preceding chapters arguing carries genuine biological importance is precisely the architecture that, when destabilized through repeat expansion, produces some of the most severe and best-documented inherited diseases in all of human medicine. Function and vulnerability, in the non-coding genome as in the coding genome this framework has traditionally centered, turn out to be two faces of the identical underlying biological reality.

CHAPTER Fourteen

Reading the Dark Genome With New Eyes

FOURTEEN

How do you test whether a single letter of DNA, sitting somewhere among three billion others, actually does anything at all? For most of the history of genetics, the honest answer was: slowly, expensively, and one letter at a time. This chapter is about the technologies that changed that answer, most of them developed within roughly the past fifteen years, and about what they have already revealed when finally pointed, systematically and at scale, toward the non-coding genome this book has spent thirteen chapters exploring.

The oldest and most direct way to test whether a piece of DNA matters is simply to remove it and see what happens. This approach, called a knockout experiment when applied to an entire gene, has been a staple of genetics research for decades, but applying it systematically to non-coding regulatory elements, of which the human genome contains hundreds of thousands of candidates, was simply not practical using the gene editing technology available before around 2012. Older gene editing methods, including zinc finger nucleases and TALENs, could achieve targeted DNA modification, but each individual edit required substantial custom protein engineering, making large-scale, systematic testing across thousands of candidate regulatory elements prohibitively expensive and time-consuming for any single research group to attempt.

CRISPR-Cas9 gene editing technology, adapted for use in mammalian cells starting in 2012 and 2013 through pioneering work from several laboratories, including those of Jennifer Doudna and Emmanuelle Charpentier, whose foundational contributions to the technology earned them the 2020 Nobel Prize in Chemistry, changed this calculation dramatically. Because CRISPR-Cas9 targets DNA using a comparatively simple, easily reprogrammed short RNA guide sequence rather than requiring extensive custom protein engineering for every new target, researchers could suddenly design and generate thousands of distinct targeting guides relatively quickly and cheaply, opening the door to genuinely systematic, large-scale functional screening of non-coding DNA in a way that had simply been impractical before.

This capability gave rise to a specific experimental approach called CRISPR screening, in which researchers introduce thousands or even hundreds of thousands of distinct CRISPR guide sequences into a population of cells simultaneously, each guide targeting a different candidate genomic region, and then measure which specific edits produce a detectable change in some measurable cellular property, commonly cell growth rate, or the expression level of a specific gene of interest. Because each cell in the experiment typically receives only one guide sequence, and because researchers can subsequently sequence which guides are enriched or depleted among cells showing the measured phenotype, this approach allows the functional testing of enormous numbers of candidate non-coding elements within a single, comparatively affordable experiment, a scale of testing that would have required years of dedicated, gene-by-gene research effort using older experimental methods.

Applied specifically to non-coding regulatory regions, CRISPR screening studies have produced some of the most direct, unambiguous evidence yet assembled that specific enhancer elements, the regulatory sequences described in an earlier chapter of this book, are genuinely required for normal expression of their target genes, moving well beyond the correlative biochemical evidence that ENCODE's original 2012 survey had relied upon. Landmark studies published through the mid-2010s, including systematic CRISPR-based enhancer screens targeting regions surrounding well-studied genes involved in blood cell development and cancer biology, directly demonstrated that deleting specific individual enhancer elements, sometimes located tens of thousands of base pairs away from their target gene, produced measurable, specific reductions in that target gene's expression, providing exactly the kind of direct causal proof of function that comparative genomics and biochemical activity measurements alone could not offer.

A second major technological advance, developed somewhat earlier but scaled up considerably over the same general period, involves a technique called massively parallel reporter assays, commonly abbreviated MPRA. Rather than testing regulatory function by deleting a sequence from its normal genomic location, MPRA takes a different approach: researchers synthesize thousands of short candidate regulatory sequences directly, attach each one to a reporter gene whose activity can be easily and precisely measured, typically a gene that produces a detectable fluorescent or luminescent signal, and then introduce this enormous pooled library of candidate sequences into living cells simultaneously, subsequently measuring how strongly each individual candidate sequence drives reporter gene activity. Because thousands of distinct candidate sequences can be tested within a single pooled experiment, and because each sequence can be uniquely identified through an attached DNA barcode sequenced afterward, MPRA allows researchers to directly and quantitatively measure the regulatory activity of enormous numbers of candidate enhancer and promoter sequences far more efficiently than testing each one individually would allow.

MPRA studies applied specifically to disease-associated genetic variants have proven particularly valuable for a specific, medically important reason. Genome-wide association studies, the large-scale research efforts that compare the genomes of many thousands of people to identify genetic variants statistically associated with disease risk, have consistently found that the overwhelming majority, commonly cited estimates suggest somewhere around ninety percent, of disease-associated genetic variants identified through this approach fall entirely outside protein-coding gene sequence, landing instead within the regulatory landscape this book has spent considerable space describing. This finding alone represents one of the more significant, and still underappreciated, ways in which the story this book has been telling connects directly to practical human health: the genetic variation most strongly linked to common diseases such as type two diabetes, coronary artery disease, and numerous psychiatric and autoimmune conditions is predominantly non-coding variation, precisely the category of sequence this book's opening chapters described as having been historically dismissed as unimportant.

One specific, thoroughly worked out example illustrates exactly how demanding, and ultimately how rewarding, this kind of investigation can be. Beginning with genome-wide association studies published in 2007, researchers repeatedly identified a strong, highly reproducible statistical association between obesity risk and a cluster of genetic variants located within the first intron of a gene called FTO. Because the association was so consistently reproduced across many independent studies, and because it remains, among more than eleven hundred genetic loci now linked to body mass index, the single strongest common genetic association with obesity identified to date, the FTO locus quickly became one of the most intensively studied regions in the entire field of human genetics. Individuals carrying two copies of the risk version of this variant were found, across large population studies, to weigh several kilograms more on average than individuals carrying two copies of the protective version, a substantial effect size for a single common genetic variant.

The obvious initial assumption was that the FTO gene itself must be responsible, since the associated variants sit directly within its own sequence. This assumption drove years of research directly investigating FTO's own biology, and it did in fact turn out that the FTO protein performs a genuine, independently important molecular function, acting as an enzyme that chemically modifies specific bases within RNA molecules, a function unrelated to the obesity association that first drew attention to the gene. But repeated attempts to directly connect the obesity-associated variants to altered FTO protein activity or expression kept coming up empty. The genetic association was robust and highly reproducible. The proposed mechanism, that the variants were somehow altering FTO's own function, simply was not holding up under direct experimental scrutiny.

The actual mechanism, when it was finally worked out through a pair of landmark studies published in 2014 and 2015, turned out to involve a gene located nowhere near where anyone had been looking. Researchers led by Marcelo Nobrega at the University of Chicago, publishing in Nature in 2014, used chromosome conformation capture, the same three-dimensional genome mapping technique described earlier in this book, to demonstrate that the obesity-associated region within FTO's first intron physically loops across a genomic distance of approximately five hundred thousand base pairs to directly contact the promoter of an entirely different gene, called IRX3, functioning as a long-range enhancer for IRX3 rather than exerting any meaningful regulatory effect on FTO itself. Mice genetically engineered to lack IRX3 entirely were found to weigh approximately thirty percent less than normal mice, carrying substantially less body fat despite eating normally and exercising at typical levels, strong direct evidence that IRX3, not FTO, was the actual functional target of the obesity-associated genetic variation.

A second study, led by Melina Claussnitzer and published in the New England Journal of Medicine in 2015, refined the mechanism down to the level of a single genetic letter. The specific variant responsible, designated rs1421085, was shown to disrupt a short sequence motif that a repressor protein called ARID5B normally binds in order to suppress the FTO intronic enhancer's activity. In individuals carrying the protective version of this single letter, ARID5B binds successfully and keeps the enhancer quiet. In individuals carrying the risk version, a single-letter change disrupts ARID5B binding, releasing the enhancer from its normal repression and increasing its activity roughly two and a half fold in human fat cell precursors, which in turn increases expression of IRX3, and a related gene called IRX5, specifically during an early developmental window in adipocyte formation, ultimately shifting the balance of fat cell development toward energy-storing white fat and away from energy-burning processes.

The FTO story is worth dwelling on precisely because of how much painstaking work it required, and how thoroughly it illustrates the practical necessity of the functional genomics tools described throughout this chapter. A purely statistical genome-wide association study correctly and reproducibly flagged this genomic region as important to obesity risk as early as 2007. Definitively explaining why, tracing the actual causal chain from a single non-coding DNA letter, through a disrupted repressor binding site, through an activated distant enhancer, through altered expression of a gene located half a million base pairs away, took the combined efforts of multiple research groups using chromosome conformation capture, luciferase reporter assays, mouse knockout models, and CRISPR-based genome editing, across roughly eight additional years of dedicated research. This is the general pattern researchers now expect when investigating non-coding disease associations: the statistical signal is often the easy part. The mechanistic explanation underneath it, precisely the kind of explanation this book has tried to provide for the specific examples covered in its preceding chapters, typically requires years of additional, painstaking functional work, one locus at a time, using exactly the experimental toolkit this chapter has described.

The practical challenge this creates for medical researchers is substantial. A disease-associated genetic variant located within a protein-coding gene sequence often allows researchers to reasonably predict, based on how that specific mutation would alter the resulting protein's structure, what biological consequence the variant might produce. A disease-associated variant located within a non-coding regulatory region offers no such straightforward interpretation; researchers typically cannot easily predict, from sequence alone, which gene a given non-coding variant might be regulating, whether the variant actually disrupts that regulatory function at all, or what the resulting biological consequence might be. MPRA and CRISPR-based functional screening have become essential tools precisely because they allow researchers to move beyond statistical association alone and directly test, experimentally, whether a specific disease-associated non-coding variant actually alters regulatory activity in a measurable, mechanistically interpretable way, transforming a merely statistical genetic association into a concrete, testable biological hypothesis about how that specific variant contributes to disease risk.

A third and considerably more recent development involves the application of large-scale machine learning models specifically trained to predict genomic function directly from DNA sequence, representing a computational rather than purely experimental approach to the same underlying challenge. Models such as Enformer, published by a DeepMind-affiliated research team in 2021 in the journal Nature Methods, use deep learning architectures capable of processing very long stretches of DNA sequence, up to roughly two hundred thousand base pairs in Enformer's case, considerably longer than earlier computational models could handle, to predict gene expression levels and various other genomic properties directly from raw sequence input alone, without requiring the kind of direct experimental measurement that CRISPR screening and MPRA depend upon. These computational models are trained on the enormous datasets already generated by ENCODE and comparable large-scale genomic consortia, learning to recognize the complex sequence patterns associated with functional regulatory elements well enough to generate reasonably accurate predictions for entirely new, previously untested genomic regions.

It is worth being appropriately careful about what these computational models can and cannot currently offer, since this represents a genuinely fast-moving area of research where overstated claims are common in less careful popular treatments. These models generate predictions, not proof; their outputs require experimental validation, typically through the CRISPR-based and MPRA approaches described earlier in this chapter, before a specific predicted regulatory function can be considered scientifically established with confidence. What these computational approaches offer, and offer genuinely well, is a way to prioritize which of the enormous number of candidate non-coding elements across the genome are most worth the considerable time and expense of direct experimental testing, functioning as an increasingly sophisticated triage system for directing limited experimental resources toward the candidates most likely to yield genuinely important functional findings.

Taken together, these three broad categories of newer functional genomics tools, CRISPR-based screening, massively parallel reporter assays, and machine learning sequence prediction models, represent a qualitatively different phase of research into the non-coding genome compared to everything this book has described in its preceding chapters. ENCODE's original 2012 survey, for all its ambition and scale, primarily measured biochemical correlation: where transcription factors bind, where chromatin appears accessible, where histone modifications associated with regulatory activity are present. The tools described in this chapter measure something closer to direct causation: what happens, specifically and measurably, when a candidate regulatory sequence is removed, or tested in isolation, or predicted computationally from sequence pattern alone, then checked against reality. This shift from correlation toward direct, testable causation represents the genuine current frontier of research into the ninety eight percent of the genome this book is devoted to explaining, and it is actively reshaping, year by year, exactly how confidently researchers can now speak about which specific non-coding sequences matter, to which specific genes, in which specific tissues, and why.

CHAPTER Fifteen

What the Ninety Eight Percent Means for You

You began this book with a number: ninety-eight percent, give or take, depending on exactly how the count is drawn, the fraction of your genome that does not code for protein. Fourteen chapters later, that number should look considerably less like an absence and considerably more like an enormous, only partially mapped territory, one that geneticists are still actively exploring using tools that did not exist when most adults reading this book were in school.

Let's take stock of where the journey has actually led, because the specific findings covered across this book do add up to something larger than the sum of their individual stories, and it is worth naming that larger picture plainly before this book closes. The non-coding genome is not one thing. It is a coalition of genuinely distinct biological categories, each operating through its own specific mechanism, each with its own discovery history, and each with its own, sometimes quite different, relationship to human health and disease.

Transposable elements, the Alu and LINE-1 sequences covered early in this book, together account for roughly half of your total DNA, the accumulated residue of tens of millions of years of genetic parasitism, occasionally still active, occasionally disease-causing, and repeatedly, across evolutionary time, repurposed by the host genome for its own regulatory benefit. Regulatory elements, the enhancers, promoters, and silencers explored through their three-dimensional chromatin architecture, constitute the decision-making layer that allows one genome to build both a neuron and a skin cell, and turn out to house the overwhelming majority of genetic variation associated with common human disease. Non-coding RNA molecules, from the chromosome-silencing reach of XIST to the cancer-associated scaffolding of HOTAIR to the genome-wide fine-tuning performed by more than two thousand distinct microRNAs, demonstrate that RNA never intended to become protein can nonetheless perform essential, specific, mechanistically well-characterized biological work. Introns, pseudogenes, telomeres, and centromeres each carry their own distinct functional stories, ranging from enabling the vast protein diversity generated through alternative splicing to physically holding your chromosomes together correctly across billions of cell divisions.

And repeat expansion diseases, the subject of this book's most sobering chapter, demonstrate unambiguously that this same non-coding territory carries genuine capacity for serious harm when its normal architecture breaks down, a reminder that function and vulnerability are inseparable properties of the same underlying biology, not opposing categories that sort neatly into "important" and "unimportant" genomic regions.

What does this expanded understanding actually mean, practically, for the reader closing this book and returning to ordinary life? It means several concrete things, some already well underway in current medical practice, others still emerging from active research laboratories toward eventual clinical application.

It means that when researchers today investigate the genetic basis of a common disease, whether type two diabetes, coronary artery disease, schizophrenia, or dozens of other conditions with a meaningful inherited component, they now know, based on the accumulated evidence this book has walked through, to look well beyond protein-coding genes alone. The genome-wide association studies mentioned in the previous chapter have made this unavoidable: the genetic variation most strongly linked to common disease risk sits predominantly within the regulatory landscape this book has spent considerable space explaining, and any serious attempt to understand the genetic architecture of common disease now requires taking that landscape seriously as a primary subject of investigation rather than a secondary afterthought.

It means that the rapidly growing field of RNA-based therapeutics, the broad category of medicine that includes the messenger RNA vaccine technology that reached an extraordinary number of people worldwide during the coronavirus pandemic, along with the antisense oligonucleotide and small interfering RNA approaches already approved for treating certain rare genetic diseases and under active investigation for many more, draws directly on decades of basic, unglamorous research into how non-coding RNA molecules function inside human cells, research that, for much of its history, attracted only modest funding and attention precisely because the molecules involved were not expected, under the older junk DNA framework, to matter very much.

It means that cancer researchers now routinely scan non-coding regulatory regions, not just protein-coding genes, when searching for the specific genetic changes that allow a normal cell to become cancerous, because a growing, well-documented list of specific cancer-driving mutations, including mutations disrupting enhancer function and mutations reactivating telomerase, occur entirely within non-coding sequence and would have been invisible to a genetic screening approach that looked only at protein-coding genes.

It means that gene therapy, the broad field aimed at treating disease by directly correcting or compensating for problematic genetic sequence, increasingly borrows directly from the biology of transposable elements this book covered early on, since several gene therapy delivery approaches use engineered versions of retroviral or transposon-derived machinery, adapted from the same broad family of molecular tools that LINE-1 elements and their ancient viral relatives originally evolved for their own replicative purposes, to insert therapeutic genetic material into patient cells.

And it means, more broadly and perhaps most importantly for a general reader without a direct professional stake in genomics research, that the basic mental model many of us carry regarding how our own genetic code actually works deserves meaningful updating. If you learned, at some point in your education, that genes are the meaningful part of your DNA and that everything else is largely inert filler, that model was a reasonable approximation of scientific understanding several decades ago, but it has been substantially, specifically, and repeatedly revised by the accumulated research this book has traced, chapter by chapter, discovery by discovery, scientist by scientist.

None of this means every remaining question has been answered, and this book has tried, wherever the evidence genuinely remains contested or incomplete, to say so plainly rather than manufacturing false certainty for the sake of a cleaner narrative. The precise functional fraction of the non-coding genome remains genuinely debated, sitting somewhere in the wide gap between ENCODE's biochemical activity estimates and the more conservative figures derived from evolutionary conservation analysis. The specific biological purpose of the majority of individual long non-coding RNA genes remains incompletely characterized, notwithstanding the well-documented exceptions like XIST and HOTAIR this book examined closely. The relative contribution of LINE-1 activity to neurological aging and disease remains an active area of investigation rather than a fully settled question. Honest science, and this book has tried throughout to model honest science rather than a triumphant, oversimplified narrative, requires holding these open questions clearly in view rather than resolving them prematurely for narrative convenience.

What can be said with confidence, resting on the specific, well-documented, peer-reviewed evidence this book has walked through across fourteen preceding chapters, is that the direction of travel across more than five decades of accumulating research has been remarkably consistent. Susumu Ohno's careful, mathematically grounded 1972 hypothesis about mutational load has not been vindicated in its broadest popular interpretation. Barbara McClintock's decades-ahead-of-its-time insistence that mobile genetic elements were active, consequential, and worth taking seriously has been vindicated many times over, across a body of subsequent research she did not live to see in its fullest form. Discovery after discovery, across essentially every category of non-coding sequence this book has examined, has moved in the same direction: toward more documented function, not less, in territory that earlier generations of geneticists had largely agreed to stop looking at closely.

The three billion letters of DNA inside nearly every cell in your body are not, it turns out, mostly silent filler surrounding a small, meaningful core of genes. They are more accurately understood as a single, integrated system, in which the twenty thousand or so protein-coding genes represent the parts list, and the vast remaining majority represents the wiring diagram, the assembly instructions, the quality control system, and, in the language this book's title borrowed at the outset, the parts of the text that were never actually silent at all, merely unread, for the better part of a century, by a field that had not yet developed the tools or the imagination to properly hear what they were saying.

That work of listening, as the chapters of this book have tried to demonstrate, is now well underway, conducted in laboratories across the world, using tools that grow more powerful and more precise with each passing year. It began with a Nobel-winning botanist walking rows of corn on Long Island, convinced that something important was moving within genomes that nobody else could yet see. It continues today in laboratories using CRISPR screens and machine learning models to test, sequence by sequence, exactly what the rest of your genetic code has been quietly doing all along. The story this book has told is, by any reasonable measure, still being written. But the opening chapters of that story, the ones documented across the pages you have just finished reading, are no longer in serious scientific doubt. The silence inside your DNA was never really silence. It was simply a conversation that took the rest of us considerably longer to learn how to hear.

Epilogue

A final thought, offered briefly, now that the specific chapters of this book have had their say.

Somewhere in the writing of a book like this one, a particular kind of question tends to surface, whether or not the author addresses it directly on the page. If ninety-eight percent of your genetic code was so casually written off for the better part of a century, what else, in what other field entirely, might currently be sitting in a comparable position: technically documented, quietly present, and simply not yet properly understood because nobody has pointed the right instrument at it, or asked the right question about it, or been willing to spend thirty years defending an unfashionable idea the way Barbara McClintock did.

That is not a question this book can answer, and it would overreach to pretend otherwise. But it seems worth leaving with the reader, because the specific scientific history recounted across these chapters carries a lesson that travels well beyond genetics. The label "junk" was never a finding. It was a placeholder, offered honestly by careful scientists working with the tools and evidence available to them at the time, and honestly revised, piece by piece, once better tools and better evidence arrived. The three billion letters inside your cells turned out to be considerably more talkative than anyone suspected in 1972. It seems reasonable to stay curious about what else might currently be waiting for someone patient enough to listen properly.

There is a specific kind of patience this story asks readers to notice, because it does not fit the usual narrative shape of a scientific breakthrough. Most popular accounts of science favor the single decisive moment: the experiment that settles the question, the paper that changes everything overnight. The actual history recounted in this book rewards a different kind of attention. Susumu Ohno's hypothesis, Barbara McClintock's cornfields, the ENCODE consortium's decade of coordinated data collection, the years separating Ambros and Ruvkun's original 1993 paper from its 2024 Nobel recognition, the eight additional years of chromosome conformation capture and reporter assays required to fully explain the FTO locus after its statistical association was first identified in 2007: none of these arrived as a single dramatic reveal. Each involved long stretches of unglamorous, incremental, frequently uncertain work, conducted by researchers who could not have known in advance which of their careful measurements would eventually matter. If this book has a final, practical lesson to offer beyond the specific biology it has covered, it may be this: the most consequential scientific revisions rarely announce themselves as revolutions while they are happening. They accumulate, one well documented finding at a time, until the accumulated weight of evidence simply becomes impossible to keep ignoring. That is precisely what happened to the non-coding genome, and there is no particular reason to assume the process has finished.

Glossary

Alu element. A short, roughly three hundred letter non-coding DNA sequence belonging to the SINE family of transposable elements, present in more than one million copies across the human genome and accounting for approximately ten to eleven percent of total human DNA.

Alternative splicing. The process by which a single gene produces multiple distinct messenger RNA molecules, and therefore multiple distinct protein products, by including or excluding different combinations of exons during RNA processing.

Centromere. A specific region of a chromosome, built largely from repetitive alpha satellite DNA, that serves as the attachment point for the cellular machinery responsible for separating duplicated chromosomes during cell division.

Chromatin. The combined complex of DNA and associated proteins, primarily histones, that make up a chromosome, whose physical folding and compaction state helps regulate which genes are active.

CRISPR-Cas9. A gene editing technology, adapted for use in living cells beginning around 2012, that allows researchers to make targeted, precise modifications to specific DNA sequences using a programmable RNA guide molecule.

ENCODE. The Encyclopedia of DNA Elements, a large, multi-institution research consortium launched in 2003 that systematically mapped biochemical activity across the human genome, publishing landmark findings in 2012.

Enhancer. A regulatory DNA sequence, often located far from the gene it controls along the linear chromosome, that increases the activity of that gene's promoter, typically by being brought into physical proximity through three-dimensional chromatin folding.

Exon. A segment of a gene that is retained in the final, mature messenger RNA molecule after splicing and that contributes to the protein-coding sequence.

Exaptation. The evolutionary process through which a genetic sequence originally shaped for one purpose is co-opted, over evolutionary time, to serve an entirely different, often beneficial function for the host organism.

Genome-wide association study. A research method that scans the genomes of large numbers of people to identify specific genetic variants statistically associated with a particular disease or trait.

Intron. A segment of a gene that is transcribed into RNA but subsequently removed through splicing before the mature messenger RNA is translated into protein.

LINE-1. Long Interspersed Nuclear Element 1, the only fully autonomous transposable element still active in the human genome, accounting for approximately seventeen to twenty percent of total human DNA.

Long non-coding RNA. An RNA molecule longer than two hundred nucleotides that does not code for protein but instead performs a direct regulatory or structural function, such as XIST or HOTAIR.

MicroRNA. A short RNA molecule, roughly twenty two nucleotides long, that regulates gene expression by binding to complementary sequences within target messenger RNA molecules, typically blocking translation or triggering the target's degradation.

Non-coding DNA. Any DNA sequence that does not directly code for a protein, encompassing regulatory elements, non-coding RNA genes, introns, transposable elements, and structural sequence such as telomeres and centromeres.

Promoter. A DNA sequence located immediately upstream of a gene where the cellular transcription machinery assembles before transcription of that gene can begin.

Pseudogene. A DNA sequence closely resembling a functional gene but that has lost the ability to produce a normal, functional protein, typically arising through gene duplication or transposable element mediated retrotransposition.

Purifying selection. An evolutionary process through which natural selection removes harmful mutations from a population, leaving detectable evidence of sequence conservation at functionally important genomic regions.

Repeat expansion. An abnormal increase in the number of times a short DNA sequence repeats in direct succession, responsible for diseases including Huntington's disease, Fragile X syndrome, and certain forms of ALS and frontotemporal dementia.

Retrotransposon. A transposable element that moves to a new genomic location through an RNA intermediate, using reverse transcriptase to convert that RNA copy back into DNA before insertion.

Silencer. A regulatory DNA sequence that reduces or blocks the activity of a gene's promoter, functioning in the opposite direction from an enhancer.

Telomerase. An enzyme that adds new copies of repetitive telomeric sequence to the ends of chromosomes, counteracting the gradual shortening that would otherwise occur with each cell division.

Telomere. A repetitive, non-coding DNA sequence capping the ends of chromosomes, providing an expendable protective buffer against the progressive sequence loss inherent to DNA replication.

Transposable element. A DNA sequence capable of moving, or historically capable of moving, to a new location within the genome, first described by Barbara McClintock in the 1940s and 1950s.

Selected Further Reading

ENCODE Project Consortium. An integrated encyclopedia of DNA elements in the human genome. Nature, 2012.

Ohno, Susumu. So much junk DNA in our genome. In Evolution of Genetic Systems, Brookhaven Symposia in Biology, 1972.

Graur, Dan, et al. On the immortality of television sets: function in the human genome according to the evolution free gospel of ENCODE. Genome Biology and Evolution, 2013.

McClintock, Barbara. The significance of responses of the genome to challenge. Nobel Lecture, 1983.

Lander, Eric S., et al. Initial sequencing and analysis of the human genome. Nature, 2001.

Batzer, Mark A., and Deininger, Prescott L. Alu repeats and human genomic diversity. Nature Reviews Genetics, 2002.

Beck, Christine R., et al. LINE-1 retrotransposition activity in human genomes. Cell, 2010.

Brown, Carolyn J., et al. A gene from the region of the human X inactivation centre is expressed exclusively from the inactive X chromosome. Nature, 1991.

Rinn, John L., et al. Functional demarcation of active and silent chromatin domains in human HOX loci by noncoding RNAs. Cell, 2007.

Lee, Rosalind C., Feinbaum, Rhonda L., and Ambros, Victor. The C. elegans heterochronic gene lin-4 encodes small RNAs with antisense complementarity to lin-14. Cell, 1993.

Reinhart, Brenda J., et al. The 21 nucleotide let-7 RNA regulates developmental timing in Caenorhabditis elegans. Nature, 2000.

Poliseno, Laura, et al. A coding independent function of gene and pseudogene mRNAs regulates tumour biology. Nature, 2010.

Blackburn, Elizabeth H., and Gall, Joseph G. A tandemly repeated sequence at the termini of the extrachromosomal ribosomal RNA genes in Tetrahymena. Journal of Molecular Biology, 1978.

Greider, Carol W., and Blackburn, Elizabeth H. Identification of a specific telomere terminal transferase activity in Tetrahymena extracts. Cell, 1985.

Nurk, Sergey, et al. The complete sequence of a human genome. Science, 2022.

DeJesus-Hernandez, Mariely, et al. Expanded GGGGCC hexanucleotide repeat in noncoding region of C9ORF72 causes chromosome 9p linked FTD and ALS. Neuron, 2011.

Avsec, Ziga, et al. Effective gene expression prediction from sequence by integrating long range interactions. Nature Methods, 2021.

Kim, Jinkuk, et al. Patient customized oligonucleotide therapy for a rare genetic disease. New England Journal of Medicine, 2019.

Miller, Timothy M., et al. Trial of antisense oligonucleotide tofersen for SOD1 ALS. New England Journal of Medicine, 2022.

Smemo, Scott, et al. Obesity associated variants within FTO form long range functional connections with IRX3. Nature, 2014.

Claussnitzer, Melina, et al. FTO obesity variant circuitry and adipocyte browning in humans. New England Journal of Medicine, 2015.

Prabhakar, Shyam, et al. Human specific gain of function in a developmental enhancer. Science, 2008.

National Human Genome Research Institute. ENCODE: Encyclopedia of DNA Elements. Available at genome.gov and encodeproject.org.

About the Author

Ingrid Falkenrath is a science writer whose work focuses on translating overlooked corners of modern biology for general readers. Drawing on years spent following genomics research as it moved from specialist journals into mainstream medicine, she writes with a particular interest in scientific ideas that took decades to earn the recognition they deserved. The Silence Inside DNA is her latest exploration of that theme.

www.ingramcontent.com/pod-product-compliance
Lightning Source LLC
LaVergne TN
LVHW100143061126
851857LV00010B/222
9798235595095